GRAVITY
AND
GRAVITATION

Also by Ron Kurtus

Tricks for Good Grades:
Strategies to Succeed in School
(Second Edition)

GRAVITY AND GRAVITATION

DERIVATIONS, EQUATIONS AND APPLICATIONS

Ron Kurtus

SfC Publishing Co.
Lake Oswego, Oregon

SfC Publishing Co.
www.sfcpublishing.com

ron.kurtus@sfcpublishing.com

Editing and illustrations: Kurtus Technologies staff

Library of Congress Control Number 2010909970

Kurtus, Ron
    Gravity and Gravitation:
    Derivations, Equations and Applications

    1. Physics 2. Physical Science

ISBN-10: 0-9767981-5-8
ISBN-13: 978-0-9767981-5-6

Dedicated to Isaac Newton, Albert Einstein and all the other great scientists who have studied this fascinating subject.

Contents

Introduction   **21**

Part 1: GRAVITY   **27**

1.1 Overview of the Force of Gravity   31

1.2 Vectors in Gravity Equations   37

1.3 Convention for Direction in Gravity Equations   43

1.4 Gravity Constant Factors   49

1.5 Equivalence Principle of Gravity   55

1.6 Mass, Weight and Gravity   59

1.7 Horizontal Motion Unaffected by Gravity   67

Part 2: Derivations of Gravity Equations   **73**

2.1 Overview of Gravity Equation Derivations   75

2.2 Derivation of Velocity-Time Equations   81

2.3 Derivation of Displacement-Time Equations   87

2.4 Derivation of Displacement-Velocity Equations   93

Part 3: Equations for Falling Objects   **99**

3.1 Overview of Equations for Falling Objects   101

3.2 Velocity Equations for Falling Objects   105

3.3 Displacement Equations for Falling Objects   111

3.4 Time Equations for Falling Objects   115

Part 4: Equations for Objects Projected Downward   121

4.1 Overview of Equations for Objects Projected Downward   123

4.2 Velocity Equations for Objects Projected Downward   127

4.3 Displacement Equations for Objects Projected Downward   133

4.4 Time Equations for Objects Projected Downward   139

Part 5: Equations for Objects Projected Upward   145

5.1 Overview of Equations for Objects Projected Upward   147

5.2 Velocity Equations for Objects Projected Upward   151

5.3 Displacement Equations for Objects Projected Upward   159

5.4 Time Equations for Objects Projected Upward   169

Part 6: Gravity Applications   179

6.1 Potential Energy of Gravity   181

6.2 Work by Gravity Against Inertia   189

6.3 Work Against Gravity and Inertia by an External Force   199

6.4 Effect of Gravity on Sideways Motion   209

6.5 Effect of Gravity on an Artillery Projectile   215

6.6 Gravity and Newton's Cannon   227

6.7 Escape Velocity from Gravity   235

6.8 Artificial Gravity   243

6.9 Center of Gravity   251

Part 7: GRAVITATION   259

7.1 Overview of Gravitation   261

Part 8: Theories of Gravitation   **267**

8.1 Overview of Theories of Gravitation   269

8.2 Law of Universal Gravitation   275

8.3 Universal Gravitation Equation   281

8.4 General Relativity Theory of Gravitation   287

8.5 Quantum Theory of Gravitation   293

8.6 Effect of Dark Matter and Dark Energy on Gravitation   299

8.7 Gravitation as a Fundamental Force   307

Part 9: Gravitation Principles   **313**

9.1 Equivalence Principles of Gravitation   315

9.2 Similarity Between Gravitation and Electrostatic Forces   323

9.3 Gravitational Speed   327

9.4 Gravitational Potential Energy   333

Part 10: Gravitation Applications   **341**

10.1 Gravitational Force Between Two Objects   343

10.2 Cavendish Experiment   349

10.3 Influence of Gravitation in the Universe   357

10.4 Gravitation Causes Tides on Earth   363

Part 11: Center of Mass   **371**

11.1 Overview of Gravitation and Center of Mass   373

11.2 Center of Mass Definitions   379

11.3 Center of Mass Location and Motion   387

11.4 Relative Motion and Center of Mass   393

11.5 Center of Mass Components   397

11.6 Center of Mass and Radial Gravitational Motion   403

11.7 Center of Mass and Tangential Gravitational Motion   411

Part 12: Orbital Motion   **419**

12.1 Derivation of Circular Orbits Around Center of Mass   421

12.2 Orbital Motion Relative to Other Object   431

12.3 Direction Convention for Gravitational Motion   437

12.4 Circular Planetary Orbits   443

12.5 Length of Year for Planets in Gravitational Orbit   453

12.6 Effect of Velocity on Orbital Motion   463

Part 13: Escape Velocity   **471**

13.1 Overview of Gravitational Escape Velocity   473

13.2 Gravitational Escape Velocity Derivation   483

13.3 Gravitational Escape Velocity with Saturn V Rocket   491

13.4 Effect of Sun on Escape Velocity from Earth   499

13.5 Gravitational Escape Velocity for a Black Hole   505

Summary   **511**

Resources   **513**

Index   **523**

About the Author   **529**

School for Champions   **531**

Illustrations

Introduction

Part 1: GRAVITY

1.1 Overview of the Force of Gravity

1.2 Vectors in Gravity Equations

Gravity vector   38

Perpendicular components of velocity at a negative angle   39

1.3 Convention for Direction in Gravity Equations

Positive and negative displacement vectors   45

Positive and negative velocity vectors   46

Positive velocity from negative displacement   46

1.4 Gravity Constant Factors

Height or altitude above Earth's surface   52

1.5 Equivalence Principle of Gravity

Balls of different mass fall at the same rate   56

1.6 Mass, Weight and Gravity

Balance scale used to compare mass   60

Released spring accelerates mass   62

Spring stretches according to weight   64

1.7 Horizontal Motion Unaffected by Gravity

Velocity from gravity independent of horizontal velocity   68

Perpendicular components of velocity at an angle   69

Horizontal and vertical components of initial velocity   70

Angle between gravity and sideways velocity changes   71

Part 2: Derivations of Gravity Equations

2.1 Overview of Gravity Equation Derivations

2.2 Derivation of Velocity-Time Equations

Velocity-time relationship   82

2.3 Derivation of Displacement-Time Equations

Displacement-time relationship   88

2.4 Derivation of Displacement-Velocity Equations

Displacement-velocity relationship   94

Part 3: Equations for Falling Objects

3.1 Overview of Equations for Falling Objects

3.2 Velocity Equations for Falling Objects

Velocity of falling object as function of time or displacement   106

3.3 Displacement Equations for Falling Objects

Displacement of falling object as function of velocity or time   112

3.4 Time Equations for Falling Objects

Elapsed time as a function of velocity or displacement   117

Part 4: Equations for Objects Projected Downward

4.1 Overview of Equations for Objects Projected Downward

4.2 Velocity Equations for Objects Projected Downward

Velocity as a function of displacement or time   128

4.3 Displacement Equations for Objects Projected Downward

Downward displacement as a function of velocity and time   134

4.4 Time Equations for Objects Projected Downward

Time as a function of velocity and displacement   140

Part 5: Equations for Objects Projected Upward

5.1 Overview of Equations for Objects Projected Upward

5.2 Velocity Equations for Objects Projected Upward

Velocities of object projected upward at different times   153

Velocities for displacements of object projected upward   157

5.3 Displacement Equations for Objects Projected Upward

Displacements for velocities of object projected upward   161

Displacements for various times of object projected upward   164

Total distances of object projected upward   166

5.4 Time Equations for Objects Projected Upward

Times for various velocities of object projected upward   171

Times for various displacements in upward direction   175

Times for various displacements in downward direction   177

Part 6: Gravity Applications

6.1 Potential Energy of Gravity

Initial and final PE and KE   183

6.2 Work by Gravity Against Inertia

Work by gravity against inertia   191

Work as change in potential energy   193

Work as change in kinetic energy   194

6.3 Work Against Gravity and Inertia by an External Force

Work against gravity and inertia   202

6.4 Effect of Gravity on Sideways Motion

Gravity pull is same for moving and stationary objects   210

Ball thrown sideways falls at same rate as dropped ball   210

Dropped bullet and shot bullet hit ground at same time   212

6.5 Effect of Gravity on an Artillery Projectile

Projectile leaves cannon at angle θ with the ground   216

Path of projectile fired from cannon   217

Same displacement for complementary angles   222

6.6 Gravity and Newton's Cannon

Cannonball follows elliptical path to hit the ground   229

Cannonball goes into a circular orbit   230

Cannonball goes into large elliptical orbit around Earth   232

6.7 Escape Velocity from Gravity

Factors for escape velocity from Earth's gravity   237

6.8 Artificial Gravity

Rotating space station creates artificial gravity   245

6.9 Center of Gravity

Calculating CG of weights   252

Measuring CG with plumb line   253

Weight balances on sharp edge   254

Object tips over when CG passes pivot point   255

CG below balance point   255

Bat follows parabolic path as it spins   256

CG is best location to hit the ball   257

Part 7: GRAVITATION

7.1 Overview of Gravitation

Part 8: Theories of Gravitation

8.1 Overview of Theories of Gravitation

Sun's gravitation bends beam of light   271

8.2 Law of Universal Gravitation

Masses attract each other   276

8.3 Universal Gravitation Equation

Various points on object attract points on other object   284

Atoms are considered as points separated by a distance   285

8.4 General Relativity Theory of Gravitation

Straight lines curve toward mass   288

8.5 Quantum Theory of Gravitation

Transfer of gravitons between two molecules   295

8.6 Effect of Dark Matter and Dark Energy on Gravitation

Dark energy seems to push objects apart   303

8.7 Gravitation as a Fundamental Force

Strong force holds nucleus together   308

Part 9: Gravitation Principles

9.1 Equivalence Principles of Gravitation

Objects fall at same rate   315

Exception when objects are much different in size   318

Experiment in accelerating spaceship   320

9.2 Similarity Between Gravitation and Electrostatic Forces

9.3 Gravitational Speed

Gravitation waves responding to change in separation   329

9.4 Gravitational Potential Energy

Part 10: Gravitation Applications

10.1 Gravitational Force Between Two Objects

Attraction between Earth and Moon   344

10.2 Cavendish Experiment

Cavendish experiment to measure gravitation   350

10.3 Influence of Gravitation in the Universe

Typical spiral-shaped galaxy   358

Earth is slightly wider at equator than between poles   358

Binary stars orbiting each other   359

10.4 Gravitation Causes Tides on Earth

1 meter rise results in several meter rise in tide 364

Force from Moon pulls ocean toward it 364

Moon attracts ocean and Earth toward it 366

Subtraction of vectors results in double bulge 366

Alignment of Sun and Moon for spring tides 367

Part 11: Center of Mass

11.1 Overview of Gravitation and Center of Mass

CM between two uniform spheres 372

11.2 Center of Mass Definitions

Center of mass of sphere is at its geometric center 377

CM between two uniform spheres 379

Center of mass is at the midpoint for equal objects 380

Center of mass can be inside much larger object 381

11.3 Center of Mass Location and Motion

Locations of objects on coordinate line 386

11.4 Relative Motion and Center of Mass

11.5 Center of Mass Components

Motion of objects can be broken into components 395

Radial vector components of the two objects 397

Tangential vector components of the two objects 398

11.6 Center of Mass and Radial Gravitational Motion

Objects moving toward CM 402

Objects moving away from each other 404

Objects moving in the same direction 405

11.7 Center of Mass and Tangential Gravitational Motion

Factors in tangential motion 410

Part 12: Orbital Motion

12.1 Derivation of Circular Orbits Around Center of Mass

Factors in objects orbiting CM 421

Double stars follow same orbit around CM 424

Orbits when one object is much larger than other 425

12.2 Orbital Motion Relative to Other Object

Moon appears to orbit Earth 432

Earth appears to orbit Moon 432

12.3 Direction Convention for Gravitational Motion

Gravity coordinate system 436

Gravitation coordinate system 437

12.4 Circular Planetary Orbits

Earth orbits CM between it and the Moon 443

Earth's tangential velocity while orbiting the Sun 444

12.5 Length of Year for Planets in Gravitational Orbit

Earth and Jupiter orbit the Sun 457

12.6 Effect of Velocity on Orbital Motion

Cannonball does not have enough velocity to go into orbit 463

Rocket follows parabolic path 465

Part 13: Escape Velocity

13.1 Overview of Gravitational Escape Velocity

Rocket reaches escape velocity   473

13.2 Gravitational Escape Velocity Derivation

Factors involved in gravitational escape velocity   483

13.3 Gravitational Escape Velocity with Saturn V Rocket

Saturn V parking orbit and required escape velocity   491

13.4 Effect of Sun on Escape Velocity from Earth

Rocket leaves Earth away from Sun   499

13.5 Gravitational Escape Velocity for a Black Hole

Black Hole and Schwarzschild radius   506

Summary
Resources
Index
About the Author

Ron Kurtus   529

School for Champions

Introduction

Gravity and gravitation are fascinating subjects that affect us all. They are important parts of the study of Physics.

If you are a student taking a Physics class, the information in this book can be a valuable supplement to your existing studies. On the other hand, if you are simply a science enthusiast, the book will not only explain concepts but also enlighten you.

In either case, much of the material presents a new way of looking at the subjects.

Book in two parts

This book is divided into two major parts: the study of gravity near the Earth and the study of gravitation for objects and astronomical bodies far from Earth.

> **Note**: Although the material is divided in a logical order, you can easily skip around to areas of interest.

Gravity

We experience gravity in our everyday lives, since it is the attraction of objects—including our own bodies—toward the Earth. Falling objects provide a good subject for scientific measurements and experiments.

Gravity concepts include the fact that since the acceleration due to gravity is considered constant near the Earth, all objects fall

---

at the same rate and horizontal motion is independent of the force of gravity. It also explains the difference between mass and weight.

Derivations

As per the subtitle of the book, I have provided derivations of gravity equations for the velocity, distance and time that apply to objects dropped, thrown downward or projected upward.

These equations allow you to verify the laws of gravity, as well as to apply them to predict such things as the distance an artillery shell will travel.

Applications

Other applications of gravity concepts and equations include determining the final velocity of a falling object, sending a projectile into space, creating artificial gravity and determining how much work is required to lift an object against gravity.

Gravitation

Beyond gravity is gravitation, which is the attraction of objects of mass toward each other. When objects are relatively close to the Earth, gravitation can be approximated as gravity.

There are three major theories of what causes gravitation: *Newton's Law of Universal Gravitation*, the *General Relativity Theory of Gravitation* and the *Quantum Theory of Gravitation*. These theories are somewhat at odds with each other.

To confuse matters, there is also the question of the effect of dark matter on gravitation.

Most of our studies involve using *Newton's Universal Gravitational Equation*, which is seen to be similar to *Coulomb's Law* of electrostatic force.

Newton's equation allows you to measure the gravitational force between two objects, understand how gravitation causes tides on Earth and determine the effect of the center of mass between two objects on their motion.

Other applications of gravitation concepts and equations include examining circular gravitational orbits and determining the velocity required for an object to escape the gravitational pull of the Earth, Sun or other celestial bodies.

Knowledge needed

In order to appreciate the material in this book, you need to have a basic knowledge of matter, motion and Newton's laws in Physics or Physical Science.

You also should be familiar with Algebra for understanding equations and solving problems through substitution.

Although the derivations employ basic Calculus, you do not need knowledge of Calculus to use this book. Advanced Physics and Physical Science students are often required to derive equations. Beginning students who have not yet taken Calculus can still understand the derivations.

Conventions

There are a number of conventions we follow in this book. They are spelled out here for better understanding.

Multiplication sign

Typically, the algebraic expression **abc** denotes **a** times **b** times **c**. Sometimes—especially when there are numbers involved—the multiplication sign must be used for reasons of clarity, such as: 5×10^3.

To avoid possible confusion with the letter "x", we use the asterisk for multiplication: $5*10^3$. This follows the common usage in many scientific websites.

Exponents

Exponents are used to indicate multiplying a number times itself. Examples are: x^2 means $x*x$ and y^3 equals $y*y*y$. This is the common convention in Algebra.

The use of exponents is especially convenient when dealing with very large or very small numbers that have a series of zeros. For example, 100,000 can be written as 10^5, which means 10 times itself 5 times or 10 followed by 5 zeros. Thus, 300,000 would be $3*10^5$.

A negative exponent indicates a reciprocal or a small number. For example, $y^{-3} = 1/y^3$. Also $10^{-5} = 1/10^5 = 0.00001$.

> **Note** that a decimal less than 1 should always be written with a zero before the decimal point to avoid any possible confusion. The decimal 0.05 is the correct form, while .05 is not.

Scientific notation

Numbers that are either too large or too small to be conveniently written as decimals can be written in the scientific or exponential notation.

This method consists of writing the number as a decimal fraction less than ten, multiplied by an exponent to the base 10.

For example, the number 286,000 would be written in scientific notation as $2.86*10^5$. Likewise, 0.003178 would be correctly written as $3.178*10^{-3}$.

Significant figures

In dealing with scientific numbers, the question arises concerning how accurate you need to be. We will follow the convention of stating accuracy to three decimal places or three significant figures.

In other words, a number such as the gravitational constant

$$G = 6.67428*10^{-11}$$

would be written as

$$G = 6.674*10^{-11}$$

This is essentially defining at what point we will round off a number, such that it is accurate enough for practical purposes.

Units of measurement

It is important to include the units of measurement in the solution of equations. This is to make sure you are using the same measurement system in your calculations.

You do not want to be using one variable in the English system of measurement—such as feet—and multiply it times a measure in the metric or SI system—such as meters. You also want to make sure you follow the units within a system. You don't want one item to be in grams, while another is in kilograms.

An advantage of including units is they assure your equations are correct. Units should cancel out, giving the resulting units you want. For example, consider $d = vt$, with $v = 10$ m/s and $t = 5$ s. If you write the solution as $d = (10$ m/s$)*(5$ s$)$, you can see that multiplying (m/s)*(s) results in meters as the correct units for distance.

Denoting equations and vectors

To help them stand out from the text, equations and variables are denoted in **boldface**. But note that many textbooks only denote vectors in boldface. Since we are not concerned that much with vectors in our equations, we will simply give an indication when vectors are being used.

Scientific method

When stating equations in Physics, you must be sure there are no misunderstandings about what you mean. For that reason, I've defined what each item in an equation means, including their units or measurement, even if the information had been explained in a previous chapter.

Also, repetition of an equation or concept can help to ingrain it in your memory. The convention used is seen in the example:

$$F = mg$$

where

- **F** is the force in newtons (N)
- **m** is the mass in kilograms (kg)
- **g** is the acceleration due to gravity in m/s^2

I believe that repetition is a great way to learn. It also helps to keep concepts clearly defined.

Careless science

One major peeve I have is the careless use of expressions and concepts that is seen in many science classes and textbooks.

One example is the difference between the concepts of gravity and gravitation. Another example is the idea that an object has an instantaneous velocity, as opposed to accelerating up to that velocity.

One reason for this careless science is the attitude that "You should have learned that last year" or "You should be able to figure it out yourself." Another reason is that perhaps the teachers, writers or scientists never really thought about it.

Of course, I'm not perfect, and I am sure I've made some mistakes in this book. However, I try to set a good example for my readers in being the best you can be.

Mini-quizzes

In order to facilitate learning, I've included a short quiz at the end of each chapter. There are only three questions, but they should help reinforce your understanding.

Conclusion

Enjoy the book. Learn from it. Ask questions.

Part 1: GRAVITY

Gravity is the attraction of objects toward the Earth, provided they are much smaller than the Earth and not too far away from the ground. It is an approximation of gravitation which applies to objects and astronomical bodies far from Earth.

Parts 1 - 6 of this book concern the equations and applications of gravity.

Part 1 Chapters

Chapters in Part 1 of this book include:

1.1 Overview of the Force of Gravity

This chapter gives a brief overview of the gravity equation, major characteristic of gravity and how gravity applies elsewhere.

1.2 Vectors in Gravity Equations

The acceleration due to gravity, velocity and displacement are vectors with magnitude and direction. This chapter also discusses scalar quantities, such as mass and time.

1.3 Convention for Direction in Gravity Equations

This chapter defines the direction of vectors used in gravity equations as being positive in the direction of gravity and those opposing gravity as negative.

1.4 Gravity Constant

The acceleration due to gravity is considered a constant for objects relatively near the Earth's surface. This chapter gives the derivation of the gravity constant, states the value of the constant at the Earth's surface and shows how gravity varies with altitude.

1.5 Equivalence Principle of Gravity

The Equivalence Principle of Gravity states that all objects fall at the same rate, assuming negligible air resistance. This chapter shows what happens when you drop two balls at the same time as a way to prove the principle.

1.6 Mass, Weight and Gravity

This chapter explains the difference between mass and weight and shows how they are measured.

1.7 Horizontal Motion Unaffected by Gravity

This chapter shows how horizontal motion or velocities perpendicular to the force of gravity is not affected by gravity. When the motion is projected at an angle, the horizontal component of the vector is unaffected. The chapter also shows that this rule is only applicable at short distances, where the curvature of the Earth is not a factor.

Other Gravity Parts

Other parts of the book on the subject of gravity include:

Part 2: Derivations of Gravity Equations

Derivations of gravity equations for velocity, displacement and time.

Part 3: Equations for Falling Objects

Equations and examples of objects falling under the influence of gravity.

Part 4: Equations for Objects Projected Downward

Equations and examples of objects projected downward under the influence of gravity.

Part 5: Equations for Objects Projected Upward

Equations and examples of objects projected upward under the influence of gravity.

Part 6: Gravity Applications

Real-world applications of the gravity equations.

1.1 Overview of the Force of Gravity

Gravity is a force that attracts objects toward the Earth. It is an approximation of the *gravitational* force that attracts objects of mass toward each other at great distances. Gravity applies to objects on or near the surface of the Earth.

The equation for the force of gravity is **F = mg**, where **g** is the acceleration due to gravity, can be designated in metric or English units. The equation also indicates the weight of an object.

The major result of this force is that all objects fall at the same rate, regardless of their mass. Gravity on the Moon and on other planets have different values of the acceleration due to gravity. However, the effects of the force are similar.

Gravity equation

According to Newton's *Law of Universal Gravitation*, gravitation is the force that attracts objects toward each other.

For objects relatively close to the Earth, this force is called *gravity*, and its equation is:

$$F = mg$$

where

- **F** is the force pulling objects toward the Earth
- **m** is the mass of the object
- **g** is the acceleration due to gravity; this number is a constant for all masses of matter
- **mg** is the product of **m** times **g**

Note: For verification that the gravitational force equals the force of gravity for objects close to Earth, see the *Gravity Constant Factors* chapter.

Acceleration due to gravity

The acceleration due to the force of gravity on Earth is designated by **g**. Its value is:

> **g** = 9.807 meters per second-squared (m/s^2) in the metric or SI system of measurement

> **g** = 32.2 feet per second-squared (ft/s^2) in the English system of measurement

> **Note**: Since most textbooks use **g** = 9.8 m/s^2 and 32 ft/s^2, we will also use the rounded-off version in these lessons.

In the equation **F** = **mg**, you must use the same measurement system for mass, **m**, as you do for **g**.

> **Note**: **g** is often incorrectly called the *acceleration **of** gravity*. That is misleading, since gravity does not accelerate. The expression should be the *acceleration **due to** gravity*, which is correct description of **g**.

Weight and mass

The weight of an object is the measurement of the force of gravity on that object:

> **w** = **mg**

where

- **w** is the weight in newtons (N) or pound-force (lb)
- **m** is the mass in kilograms or pound-mass (lb-mass)

> **Note**: There is often confusion concerning the designation of weight and mass.

> Although a kilogram is supposed to be a unit of mass, it is often used to designate weight. You must be aware that weight of 1 kg of mass is **w** = 9.8 newtons.

> Also, a pound is supposed to be a force, but is often called a mass. The mass of 1 pound-force is 1/32 pound-mass.

Weighing an object

You can find the weight of an object on a calibrated scale—usually with a spring resisting the force of the weight.

The mass of an object can be measured with a balance scale, comparing with an object of a given mass.

Objects fall at the same rate

The most outstanding characteristic of gravity is the fact that all objects fall at the same rate—assuming the effect of air resistance is negligible. This is because the acceleration due to gravity, **g**, is a constant for all objects, no matter what their mass.

It seems counterintuitive, since you would expect a heavy object to fall faster than an object that weighed less. But it is a fact.

Try dropping two objects at the same time, from the same height, making sure they are heavy enough not to be affected by air resistance. You will see they hit the ground at the same time.

Gravity elsewhere

When you talk about gravity, you mean gravitation near the Earth. However, the same gravity equation holds for objects near the Moon or other planets, except that the value of **g** is different.

In those cases, you typically tell where the gravity is, such as "gravity on the Moon" or "gravity on Mars."

Gravity on the Moon

Since the acceleration due to gravity on the Moon $\mathbf{g}_m$ is 1.6 m/s^2 or 5.3 ft/s^2, the force of gravity on the Moon is approximately 1/6 of that on the Earth for a given mass. Thus:

$$\mathbf{F}_m = m\mathbf{g}_m$$

where $\mathbf{F}_m$ is the force or weight on the Moon.

Weight on the Moon

The value for g_m is approximately 1/6 of the value for g on Earth. Thus, an object on the Moon would weigh about 1/6 of its weight on Earth.

Using a spring scale, if you weigh 60 kg (132 pounds) on the Earth, you would weight only 10 kg (22 lbs) on the Moon. However, using a balance scale on both Earth and the Moon, your mass would be the same.

Dropped objects

If you dropped two objects of different weights on the Moon, they would fall to the ground at the same rate. You wouldn't have to worry about the effect of air resistance, since there is no air on the Moon.

Since $g_m = g/6$, the objects would fall at a slower rate.

> (See chapter *3.1 Overview of Gravity Equations for Falling Objects* and then apply g_m to get the different values.)

Summary

Gravity is the force that pulls objects toward the Earth. It is a special case of gravitation. The equation for the force due to gravity is $F = mg$, resulting in the fact that all objects fall at the same rate, regardless of their mass.

Gravity on the Moon and on other planets have different values of the acceleration due to gravity, but the effects of the force are similar.

Mini-quiz to check your understanding

1. What is required for the gravity equation to apply?

   a. There are no requirements, and it always applies

   b. Objects must be relatively close to the Earth

   c. There must be no air resistance

2. Do a piece of tissue and a ball fall at the same rate?

   a. Yes, because everything falls at the same rate

   b. Scientists aren't sure, because it has never been tried before

   c. No, because air resistance slows down the tissue

3. Why would an object fall slower on the Moon?

   a. Because the acceleration due to gravity on the Moon is less than on Earth

   b. It would actually fall faster because there is no air resistance

   c. Gravity from the Earth would slow it down

Answers

1b, 2c, 3a

1.2 Vectors in Gravity Equations

A *vector* is a quantity that has magnitude and direction. Vectors are usually designated as arrows pointing in a specific direction, with the length as the magnitude. This provides a way to visualize and to combine entities geometrically.

Using vectors is an effective way to describe the motion of objects in gravity equations, as well as to set a convention for direction in both the vertical and horizontal axes.

Vectors are also useful to describe motion at an angle to the force of gravity, since the vector can be broken into its vertical and horizontal components.

If the quantity has only a magnitude and does not have a specific direction, it is a scalar quantity.

Vectors

Certain motion entities not only have magnitude but also have a direction. They can be represented geometrically as arrows or vectors.

Vectors are quantities with magnitude and direction. For example, a truck is traveling north at a velocity of 50 miles per hour. Its velocity is a vector with north as the direction and 50 miles per hour as the magnitude.

Vectors used in gravity equations include:
- **g**: Acceleration due to gravity
- **F**: Force of gravity
- **y** and **x**: Displacement
- **v**: Velocity

Note: Textbooks often denote vectors as with a special marker over the vectors as an indicator. In our material, vectors, scalars and equations are **boldfaced** to distinguish them from the other text. If an item is a vector, it will be noted as such.

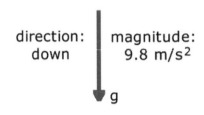

Gravity vector

Coordinate system

When dealing with vectors and gravity equations, the **x-y** or Cartesian coordinate system is used, with the **y**-axis as vertical and the **x**-axis as horizontal. The zero-point of the axis is the starting point of the object's motion.

Direction convention

The convention for direction that we use is that vectors toward the ground are positive and those upward are negative. This essentially is inverting or flipping the usual **x-y** coordinate system , such that the **−y**-axis is up instead of downward.

> (See chapter *1.3 Convention for Direction in Gravity Equations* for more information.)

Magnitude of vectors

The magnitude of a vector is always a positive number, no matter what the direction of the vector. It is an indication of the length of the geometric representation, as well as a multiplier when comparing two vectors.

Vector components

Vectors at an angle to the ground or horizontal plane can be broken into vertical and horizontal components. This includes velocity and displacement. Obviously, gravity is only in the vertical direction.

Angles

The convention for angles is that they are measured clockwise from the horizontal or **x**-axis. An angle measured counterclockwise from the **x**-axis results in a negative angle or 360° minus the angle.

Geometric visualization

Consider the velocity at an upward angle from the ground. It would be represented by a negative vector, **−v**. The angle from the horizontal would be counterclockwise or a negative angle, **−θ** (Greek letter theta).

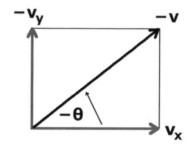

Perpendicular components of velocity at a negative angle

The velocity vector can be broken into its perpendicular components, where:

$$-v_y = -v*\sin(-\theta)$$

and

$$v_x = -v*\cos(-\theta)$$

Also, according to the Pythagorean Theorem:

$$v^2 = v_x^2 + v_y^2$$

Scalars

If the quantity only has a magnitude and does not have a direction, it is a scalar quantity. Scalar quantities can only be positive.

Scalar quantities used in gravity equations are:

- Mass: **m**
- Time: **t**
- Speed: **s**
- Distance: **d**

Speed versus velocity

Speed is how fast an object is going irrespective of its path. *Velocity* is a vector that is the speed in a specific direction.

Distance versus displacement

While *displacement* is a vector that shows how far an object moves in some direction, *distance* is the total of the path taken between two points.

Vectors as scalars

The absolute or positive value of a vector is its magnitude or scalar quantity. For example:

$$|\mathbf{v}| = \mathbf{s}$$

$$|\mathbf{y}| = \mathbf{d}$$

where

- $|\mathbf{v}|$ is the absolute or positive value of the velocity, independent of the direction
- $|\mathbf{y}|$ is the absolute value of the displacement, independent of direction

Likewise:

$$|-\mathbf{v}| = \mathbf{s}$$

$$|-\mathbf{y}| = \mathbf{d}$$

Summary

A vector is a quantity that has magnitude and direction. Using vectors is an effective way to describe the motion of objects in gravity equations. They are also are also useful to describe motion at an angle to the force of gravity, since the vector can be broken into its vertical and horizontal components.

If the quantity has only a magnitude and does not have a specific direction, it is a scalar quantity. A scalar quantity is also the absolute value of its vector.

Mini-quiz to check your understanding

1. What does the length of a vector arrow represent?

   a. The magnitude of the quantity

   b. The direction of the motion

   c. It is only used for visualization

2. When is the angle between a vector and the horizontal axis positive?

   a. The angle is always negative

   b. Whenever the Pythagorean Theorem holds

   c. When the vector is positive

3. Why is distance a scalar quantity?

   a. Distance is not a scalar but a vector

   b. Distance does not depend on direction of travel

   c. Because distance equals velocity times time

Answers

1a, 2c, 3b

1.3 Convention for Direction in Gravity Equations

Since some gravity equations result in an object changing directions, there needs to be a convention as to which direction is positive and which is negative.

Due to the fact that the force of gravity is downward, it would only seem logical to define downward as a positive direction in gravity equations. This is essentially inverting the Cartesian coordinate system.

Using vectors is an effective way to describe the initial and resulting motion of the objects and to set a convention for direction.

We consider the displacement direction as positive below the starting point and negative above. Velocity is positive in a downward direction and negative when going upward. For horizontal motion, right is positive and left is negative.

Gravity convention

The acceleration due to gravity, **g**, is in a direction toward the ground. The same is true of the force of gravity, **F = mg**.

Both **g** and **F** can be considered vectors, which are geometric representations indicating both magnitude and direction. Vectors are typically drawn as arrows, with their length proportional to their magnitude. The magnitude of **g** is 9.8 m/s^2 or 32 ft/s^2.

Since both the acceleration due to gravity and the force of gravity are downward, we use the convention that *downward vectors are positive* (+). That means that *upward vectors are negative* (−).

In essence, this is flipping the **x-y** or Cartesian coordinate system, such that the **+y**-axis points down and the **−y**-axis points up.

This convention affects the other vectors used in gravity equations, which include:

- Displacement: **y** and **x**
- Velocity: **v**

Displacement convention

Displacement is the amount of movement or change in position from the starting point in a given direction. It is a vector, with a direction of either downward, upward or at an angle with the vertical line.

> **Note**: *Displacement* is often confused with *distance*. Displacement concerns a direct path from one point to another and is a vector. Distance is a scalar quantity where the path does not matter.

Magnitude

The magnitude of displacement is the *separation* between the starting and end points along the vector line. Distance can also be considered as the magnitude of displacement, provided it is along the vector line.

Vertical displacement

Vertical displacement can be represented as a vector that is positive below the starting point and in the direction of gravity. It is negative above the starting point.

> **Note** that when an object is projected upward, reaches its maximum displacement and starts moving downward, the displacement is negative above the starting point and positive below the starting point. Displacement is measured from the starting point.

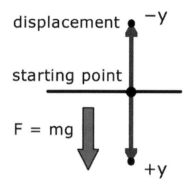

Positive and negative displacement vectors

Horizontal displacement

Displacement vectors in a horizontal direction are designated as positive toward the right direction and negative toward the left. Vectors at an angle can be broken into their components on the **x-y** axis.

Velocity convention

Velocity is a vector that is the change in displacement with respect to time in a specific direction. It is measured from a given displacement.

Velocities in the same direction as gravity are positive and velocities in the opposite direction are negative vectors.

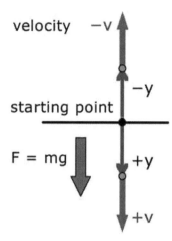

Positive and negative velocity vectors

Magnitude

Velocities in the same direction as gravity are positive and velocities in the opposite direction are negative vectors.

Velocity at a negative displacement

It is possible for a velocity at a negative displacement to be in the direction of gravity and be a positive vector. This is seen in the case of throwing a ball upward and having it fall toward the ground.

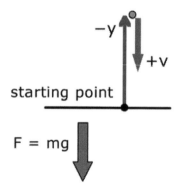

Positive velocity from negative displacement

Velocity in horizontal direction

Velocity vectors in a horizontal direction are designated as positive toward the right direction and negative toward the left. Vectors at an angle can be broken into their components on the **x-y** axes.

Summary

Gravity equations require a convention, stating which direction is positive and which is negative. Using vectors is an effective way to describe the initial and resulting motion of the objects and to set a convention for direction.

We consider the direction of gravity as positive and the upward direction as negative. Displacement is positive below the starting point and negative above. Velocity is positive in a downward direction and negative when going upward.

For horizontal motion, right is positive and left is negative.

Mini-quiz to check your understanding

1. Why isn't "up" considered positive as it is in the x-y coordinate system?

    a. There is no such thing as upward or downward in the x-y coordinate system

    b. Since the force of gravity is downward, it is more logical to have that direction as positive for gravity studies

    c. Many people have wondered about that convention

2. From where is the displacement of a projected object measured?

    a. Displacement is measured from the starting point

    b. Displacement is always measured from the ground

    c. Displacement varies according to distance traveled

3. If a ball is thrown upward, what sign is its velocity on the way down?

    a. It is negative, since it started going upward

    b. It is positive, since it is moving toward the ground

    c. It depends whether the displacement is positive or negative

Answers

1b, 2a, 3b

1.4 Gravity Constant Factors

The value of **g** in the gravity force equation **F = mg** is the acceleration due to gravity. It is considered a constant for objects relatively near the Earth's surface.

The gravity constant comes from the *Universal Gravitation Equation* at the Earth's surface. By substituting in values for the mass and radius of the Earth, you can calculate the value of the gravity constant at the Earth's surface.

The fact that the acceleration due to gravity is a constant facilitates the derivations of the gravity equations for falling objects, as well as those projected downward or upward. However, the value of **g** starts to vary at high altitudes.

Derivation of gravity constant

The acceleration due to gravity constant comes from Newton's Universal Gravitation Equation, which shows the force of attraction between any two objects—typically astronomical objects:

F = GMm/R²

where

- **F** is the force of attraction, as measured in newtons (N) or kg-m/s²
- **G** is the Universal Gravitational Constant: $6.674*10^{-11}$ m³/s²-kg
- **M** and **m** are the masses of the objects in kilograms (kg)
- **R** is the separation of the centers of the objects in meters (m)

(See chapter *8.3 Universal Gravitation Equation* for more information.)

One assumption made is that the mass of each object is concentrated at its center. Thus, if you considered a hypothetical point object of mass **m** that was at the surface of the Earth, the force between them would be:

$$F = GM_E m/R_E^2$$

where

- **F** is the force of attraction at the Earth's surface
- **G** is the Universal Gravitational Constant
- M_E is the mass of the Earth
- **m** is the mass of the object
- R_E is the separation between the center of the Earth and an object on its surface; it is also the radius of the Earth

Since GM_E/R_E^2 is a constant, set:

$$g = GM_E/R_E^2$$

This is the gravity constant or acceleration due to gravity. Thus, the gravity equation is:

$$F = mg$$

Value of g

You can find the value of **g** by substituting the following items into the $g = GM_E/R_E^2$ equation:

$G = 6.674*10^{-11}$ m^3/s^2-kg

$M_E = 5.974*10^{24}$ kg

$R_E = 6.371*10^6$ m

Note: Since the Earth is not a perfect sphere, the radius varies in different locations, including being greater at the equator and less at the poles. The accepted average or mean radius is 6371 km.

The result is:

$\mathbf{g} = (6.674*10^{-11} \text{ m}^3/\text{s}^2\text{-kg})(5.974*10^{24} \text{ kg})/ (6.371*10^6 \text{ m})^2$

$\mathbf{g} = (6.674*10^{-11})(5.974*10^{24})/(40.590*10^{12}) \text{ m/s}^2$

$\mathbf{g} = 0.9823*10^1 \text{ m/s}^2$

$\mathbf{g} = 9.823 \text{ m/s}^2$

This value is close to the official value of $\mathbf{g} = 9.807$ m/s^2 or 32.174 ft/s^2, defined by the international *General Conference on Weights and Measures* in 1901. Factors such as the rotation of the Earth and the effect of large masses of matter, such as mountains were taken into effect in their definition.

Although, the value of $\mathbf{g}$ varies from place to place around the world, we use the common values of:

$\mathbf{g} = 9.8$ m/s^2 or 32 ft/s^2

On other planets

The same principles of gravity on Earth can apply to other astronomical bodies, when objects are relatively close to the planet or moon.

We typically consider "gravity" as concerning Earth. If you are talking about the force of gravity on another planet, you should say, "gravity on Mars" or such.

Acceleration due to gravity on the:

Earth: 9.8 m/s^2

Moon: 1.6 m/s^2

Mars: 3.7 m/s^2

Sun: 275 m/s^2

Variation of gravity with altitude

Although **g** is considered a constant, its value does vary with altitude or height from the ground. You can show the variation with height from the equation:

$$g_h = GM_E/(R_E + h)^2$$

where

- g_h is the acceleration due to gravity at height **h**
- **h** is the height above the Earth's surface or the altitude of the object

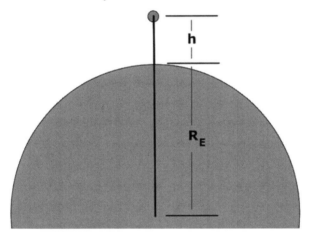

Height or altitude above Earth's surface

To facilitate calculations, it is easier to state **h** as a percentage or decimal fraction of R_E. For example, if h = 10% of R_E or $0.1R_E$, then:

$$g_h = GM_E/(1.1R_E)^2$$

and

$$g_h = 0.826GM_E/R_E{}^2 = 0.826g$$

Charting **h** and g_h:

h	%R_E	g_h
67.31 m (220.8 ft)	0.001%	$0.99998g = 9.8$ m/s^2
637.1 m (2207.8 ft)	0.01%	$0.9998g = 9.8$ m/s^2
6.371 km (3.95 mi)	0.1%	$0.998g = 9.78$ m/s^2
63.71 km (39.5 mi)	1%	$0.980g = 9.6$ m/s^2
637.1 km (395 mi)	10%	$0.826g = 8.09$ m/s^2

As you can see, the value of **g** starts to deviate from 9.8 m/s2 at about 6.4 km or 4 miles in altitude. At about 64 km or 40 mi, the change in **g** is sufficient to noticeably affect the results of gravity equations.

Effect on gravity derivations

The derivations of the equations for velocity, time and displacement for objects dropped, projected downward, or projected upward depend on **g** being a constant. Even a 1% or 2% variation in the value of **g** can affect the derivations.

> (See chapter *2.1 Overview of Gravity Equation Derivations* for more information.)

Summary

The acceleration due to gravity, **g**, is considered a constant and comes from the Universal Gravitation Equation, calculated at the Earth's surface. By substituting in values for the mass and radius of the Earth, you can find the value of **g**.

A constant acceleration due to gravity facilitates the derivations of the gravity equations. However, the value of **g** starts to vary at high altitudes.

Mini-quiz to check your understanding

1. What does the radius of the Earth signify in deriving the gravity constant?

    a. It is not a factor in the derivation

    b. It is an extremely large number that makes gravity constant

    c. It is the separation between the Earth's center and that of an object on the surface

2. What factors affect the value of gravity?

    a. The radius of the Earth varies in different locations, plus the effect of the Earth's rotation

    b. The atmosphere is thinner on mountain tops, affecting the gravity

    c. Calculating errors, the effect of the Sun on gravity and the seasons

3. What happens to **g** as you reach higher altitudes?

    a. It gets larger, approaching infinity

    b. Nothing, since **g** is a constant

    c. It becomes a smaller value

Answers

1c, 2a, 3c

1.5 Equivalence Principle of Gravity

Intuitively, you would think that a heavier object would fall to the ground faster than a lighter object. However, that is not the case.

The *Equivalence Principle of Gravity* (also called the *Weak Equivalence Principle* or *Uniqueness of Free Fall Principle*) states that all objects fall at the same rate, assuming negligible air resistance.

By examining the equation for the force of gravity, you can see that the value of the acceleration due to gravity is a constant and is independent of the mass of the object. A constant acceleration means that objects fall at the same rate. This principle has been proven experimentally many times.

Acceleration due to gravity is constant

You can see that the acceleration due to gravity is constant by examining the force of gravity on objects relatively close to Earth:

$$F = mg$$

where

- **F** is the force in newtons (N) or pound-force (lb)
- **m** is the mass of the object in kilograms (kg) or pound-mass
- **g** is the acceleration due to gravity (9.8 m/s^2 or 32 ft/s^2)

What this means is that the force from the Earth's gravity is proportional to the mass of the object. Objects with greater mass feel a greater the force on them.

However, the acceleration due to gravity, **g**, is the same for all objects, no matter what their mass. This means that all objects will accelerate or fall at the same rate, provided they are not affected by air resistance.

Balls fall at same rate

Dropping two objects of different mass from exactly the same height and exactly the same time will result in them falling at the same rate and hitting the ground simultaneously.

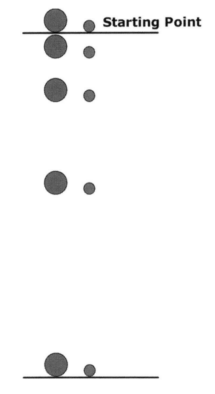

Balls of different mass fall at the same rate

If one or both objects are noticeably affected by air resistance, another factor comes into play, and the rule does not hold. For example, dropping a golf ball and a feather will result in the golf ball hitting the ground before the feather, which is greatly affected by air resistance and air currents.

Note that any object you drop is somewhat affected by air resistance. However, some are barely affected, such that the rule still holds.

Experimental verification

The fact that the acceleration due to gravity is independent of the mass of the objects has been verified many times.

In the 1600s, Galileo Galilei was said to have dropped two balls of different mass from the Leaning Tower of Pisa to prove that objects of different mass fall at the same rate.

Some historians doubt whether he actually did the experiment at Pisa, but the experimental results are documented.

You can verify this experiment yourself by standing on a chair and dropping two balls or objects of different weights at exactly the same time. This is a rough experiment, but it can demonstrate the principle.

Summary

The *Equivalence Principle of Gravity* states that all objects fall at the same rate, assuming negligible air resistance.

The force of gravity equation shows that the value of the acceleration due to gravity is a constant and is independent of the mass of the object. A constant acceleration means that objects fall at the same rate.

This principle has been proven experimentally many times. The most famous verification was when Galileo apparently dropped two balls of different mass from the Leaning Tower of Pisa.

Mini-quiz to check your understanding

1. What happens when you increase the mass of an object?

   a. The force of gravity on it increases

   b. The acceleration due to gravity increases

   c. Its weight decreases

2. When does the rule that different masses fall at the same rate fail?

   a. The rule is always true, no matter what

   b. It fails every other time you try it

   c. When an object is noticeably affected by air resistance

3. How could Galileo tell whether the balls hit the ground at the same time, after he dropped them from the Leaning Tower of Pisa?

   a. He used a stopwatch to time them

   b. He had someone standing on the ground below to make the observation

   c. He just guessed that they probably hit the ground at the same time

Answers

1a, 2c, 3b

1.6 Mass, Weight and Gravity

The mass of an object is the amount of matter it contains regardless of its volume or any forces acting on it. There are two measurements of mass: gravitational and inertial.

Gravitational mass is the mass of a body as determined by its response to the force of gravity, such as done on a balance scale. Inertial mass is the measurement of the mass of an object as measured by its resistance to acceleration.

Gravitational mass and inertial mass have been shown to be equivalent.

Weight is defined as the force of gravity on a mass. A spring scale can be used to measure weight. Although mass is the same on the Moon as it is on Earth, the weight of an object is 1/6 as much on the Moon as it is on the Earth.

Gravitational mass

The measurement of the gravitational mass of an object is done by comparing the arms of a balance scale between a unit mass and the mass of interest.

Unit of mass

The unit of mass in the metric or SI system is the kilogram (kg). In the English system, the unit of mass is the pound-mass (lb-mass) or the slug.

Originally, a kilogram was defined as the amount of matter in 1 liter (1 L) of water at the temperature of melting ice (0° C). A gram is 1 cubic centimeter (1 cc) of water at 0° C.

Presently, a metal standard is used to designate a kilogram instead of a quantity of water.

Balance scale to measure mass

A balance scale is often used to measure mass by comparing the moment arms. A unit mass (1 kg) or some known mass is used as a basis of comparison with an unknown mass.

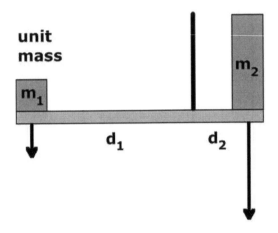

Balance scale used to compare mass

The pivot point of the scale is adjusted so that the objects balance. Then the mass is determined from the ratio:

$$m_1 d_1 = m_2 d_2$$

$$m_2 = m_1 d_1 / d_2$$

where

- m_1 is the unit mass
- d_1 is the moment arm of the unit mass
- m_2 is the test mass
- d_2 is the moment arm of the unit mass

If m_1 = 1 kg, then:

$$m_2 = d_1 / d_2 \text{ kg}$$

Although the official unit of mass is 1 kg, you may use a mass with a fraction of that mass, such as 10 grams. Likewise, the moment arm may be in centimeters, instead of meters. Similar variations are true if you are using the English system for mass and length.

Inertial mass

Inertial mass is determined by applying Newton's Second Law, which states that a force is required to accelerate a mass and overcome its inertia. This is expressed in the equation:

F = ma

where

- **F** is a force in newtons (N) or pounds (lbs) that is necessary to overcome the inertia of the mass
- **m** is the mass in kilograms (kg) or pound-mass (lbs-mass)
- **a** is the resulting acceleration in m/s^2 or ft/s^2

This mass is sometimes called inertial mass since the force is overcoming the inertia of the mass.

Using a spring to apply force

The mass of an object can be determined experimentally by applying *Hooke's Law* for springs, which says that the force is proportional to the length stretched or compressed in an ideal spring:

F = kΔx

where

- **F** is the force applied by an ideal spring
- **k** is the spring constant, which depends on the spring material
- **Δx** is the change in the length of the spring (delta **x**)

Compressing a spring requires a force. The compressed spring then has potential energy that, when released, will accelerate the mass at the free end of the spring.

Combining the force equations:

$$ma = k\Delta x$$

$$m = k\Delta x/a$$

Thus, knowing the spring constant and measuring the compression displacement and acceleration, you can determine the mass of the object. Of course, this is an ideal situation.

The mass of the spring, friction and other factors come into play.

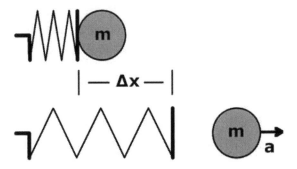

Released spring accelerates mass

Using ratio

If two different masses are accelerated by the same spring configuration, the ratio of those masses can be found.

This eliminates the need to know k and Δx.

$$m_1 a_1 = k\Delta x$$

$$m_2 a_2 = k\Delta x$$

$$m_1 a_1 = m_2 a_2$$

Also, if m_1 is a unit mass, m_2 can be found from the ratio of accelerations:

$$m_2 = a_1/a_2 \text{ kg}$$

Gravitational and inertial equivalent

Scientists have wondered if the effect of gravity on a mass was the same as the effect of acceleration. Experiments have shown that gravitational mass was equivalent to inertial mass.

This fact was instrumental in the determining of the *General Relativity Theory of Gravitation.*

Weight

The weight of an object is the force of gravity on the mass of the object:

$$F = mg$$

or

$$W = mg$$

where

- **F** is the force of gravity on the mass in newtons (N) or pounds (lbs)
- **m** is the mass of the object in kg or pound-mass
- **g** is the acceleration due to gravity (9.8 m/s^2 or 32 ft/s^2)
- **W** is the weight in N or lbs

Confusion between weight and mass

Many people (and even textbooks) mix up mass and weight. They will say that an object weighs 25 kg and another object as a mass of 2 pounds. Both expressions are scientifically incorrect.

Mass is the amount of matter. The metric or SI unit of mass is the kilogram. The English unit of mass is the slug or pound-mass.

The units of weight are units of force. The metric or SI unit of weight is the Newton (N) or sometimes the Kg-force. The English unit of weight is the pound (lb).

Measuring weight

A spring scale is often used to measure an object's weight. Again, Hooke's Law is applied. The displacement a spring is stretched for a given force or weight is first calibrated. Then you can use the spring scale to measure the weight of various objects.

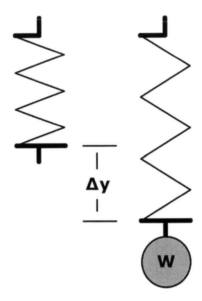

Spring stretches according to weight

Spring scales are calibrated with a known weight, such that knowing Δy will give you the weight, **W**. Within a range of weights, the weight is proportional to the displacement stretched. In other words:

$$W_2 = W_1 \Delta y_2 / \Delta y_1$$

where

- W_2 is the unknown weight
- W_1 is the unit weight or calibrating weight
- Δy_2 is the stretch of the spring for the unknown weight
- Δy_1 is the calibrating stretch of the spring

For example, if the scale had been calibrated W_1 = 1 lb stretched the spring Δy_1 = 1 inch, then if the spring was stretched

Δy_2 = 2.5 inches, W_2 = 2.5 lbs.

Weight on the Moon

The mass of an object—or the amount of matter it contains—is the same on the Moon as it is on the Earth. However, the weight of the object is a function of the acceleration due to gravity. Since gravity on the Moon is about 1/6 of that on Earth, an object will weigh 1/6 as much on the Moon.

$$W_M = mg_M$$

where

- W_M is the weight on the Moon
- m is the mass of the object
- g_M is the acceleration due to the Moon's gravity (1.6 m/s^2 or 5.3 ft/s^2)

Since $g_M = g/6$, then:

$$W_M = W/6$$

If you weigh 60 kg (132 pounds) on the Earth, you would weigh only 10 kg (22 lbs) on the Moon.

Summary

Gravitational mass of an object is determined by using a balance scale to compare its mass with a unit mass. Inertial mass is the measurement of the mass of an object measured by its resistance to acceleration.

Gravitational mass and inertial mass have been shown to be equivalent. Weight is defined as the force of gravity on a mass. A spring scale can be used to measure weight. The mass of an object is the same on the Moon as it is on Earth, but its weight is 1/6 as much on the Moon as on the Earth.

Mini-quiz to check your understanding

1. If on a balance scale, m_1 = 10 g, d_1 = 10 cm and d_2 = 5 cm, what is the mass of m_2?

   a. 100 g

   b. 20 g

   c. 5 g

2. If a known mass of 10 g is accelerated at 2 m/s, what is the mass of an object accelerated at 4 m/s on the same configuration?

   a. 5 g

   b. 10 g

   c. 20 g

3. Could a balance scale be used to measure weight?

   a. Yes, since weight is proportional to mass

   b. No, because weight requires a spring

   c. It is uncertain, since no one has ever tried it

Answers

1b, 2a, 3a

1.7 Horizontal Motion Unaffected by Gravity

The horizontal velocity of an object is unaffected by the force of gravity for relatively short displacements. This means that the horizontal velocity is constant, while the vertical velocity is accelerating.

The reason is because perpendicular vectors act independently of each other.

An object projected at an angle to gravity can be broken into its horizontal and vertical components. This horizontal velocity component is also unaffected by the force of gravity. Again, this is for short displacements.

However, for greater displacements, the curvature of the Earth comes into play, and the angle of the force of gravity changes with displacement.

This change in angle affects the motion of the object and its velocity is no longer independent.

Horizontal motion independent

For relatively short displacements, the Earth can be considered flat. In such a case, the force of gravity is continually perpendicular to an object moving in a horizontal direction.

> **Note**: *Displacement* is a vector in a specific direction. *Distance* is a scalar quanity that is not necessarily in a straight line and where no direction is indicate.
>
> (See chapter *1.2 Vectors in Gravity Equations* for more information.)

Horizontal velocity constant

From Newton's *Law of the Conservation of Momentum*, the velocity of an object moving in a given direction will remain constant provided there are no forces acting in that direction.

A force perpendicular to the direction of motion may change the direction of the object but will have no effect on the velocity in the given direction.

This means that a horizontal velocity is independent of a vertical force or resulting velocity, such as caused by gravity.

> (An application of this can be seen in chapter *6.4 Effect of Gravity on Sideways Motion.*)

In the illustration below, the initial velocity of the object (v_i) is the constant horizontal velocity (v_x). The velocity caused by gravity (v_g) is the velocity in the y-direction (v_y). The Earth is considered flat for short displacements.

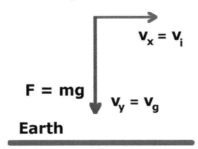

Velocity from gravity independent of horizontal velocity

Vector explanation

Another explanation is the rule that perpendicular vector quantities are independent of each other. A vector is a graphical representation of a force, acceleration, velocity or displacement, giving both magnitude and direction.

In the illustration below, v_x and v_y are perpendicular and act independently of each other. The diagonal **v** is the vector sum of the velocities and represents that actual motion of the object.

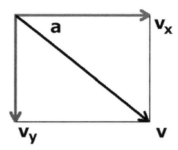

Perpendicular components of velocity at an angle

You can also break a vector into its perpendicular components. The velocity **v** in the illustration above can be broken into its **x** and **y** components.

Magnitudes

The magnitude of a velocity vector is its speed, which is the absolute or positive value of the velocity.

The magnitudes of the vectors in the illustration above are:

$$s_y = s*sin(a)$$

$$s_x = s*cos(a)$$

where

- s_y is the speed or absolute value of vector $\mathbf{v_y}$
- s is the speed for vector **v**
- **sin(a)** is the sine of angle **a**
- s_x is the absolute value of vector $\mathbf{v_x}$
- **cos(a)** is the cosine of angle **a**

Also, according to the Pythagorean Theorem:

$$s^2 = s_x{}^2 + s_y{}^2$$

(See chapter *1.2 Vectors in Gravity Equations* for more information.)

Applied to motion projected at an angle

This principle can be applied to motion when the initial velocity is projected at an angle to the ground. The motion can be broken into its horizontal and vertical components.

The vertical component of the velocity is in the same plane at gravity and is thus affected by gravity. Meanwhile the horizontal component is perpendicular to gravity and is independent of that force.

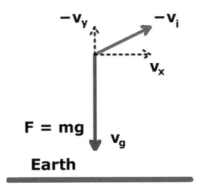

Horizontal and vertical components of initial velocity

The in illustration above, the initial velocity $-v_i$ is broken into its components in the **x** and **y** directions. According to our direction convention, up is negative.

The vertical component of the initial velocity, $-v_y$, acts only on the velocity toward the Earth, as seen in the *Velocity Equations for Objects Projected Upward* chapter. The horizontal velocity is unaffected by the vertical motion.

An example of this type of motion is seen in chapter *6.4 Effect of Gravity on an Artillery Projectile* chapter.

Situation of curved surface

The rule that perpendicular motion is unaffected by gravity is only applicable when the displacements are small enough that the Earth's surface is considered flat.

When the curvature of the Earth comes into play, the angle between the force of gravity and the motion of the object changes as the object moves. This invalidates the rule and complicates the situation. The effect of gravity now changes the horizontal velocity.

The only time the velocity remains constant is the special case of a circular orbit. In this case, the object moves at a constant velocity tangent to the circular path.

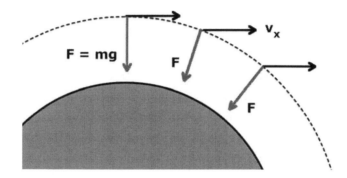

Angle between gravity and sideways velocity changes

(See chapters *6.6 Gravity and Newton's Cannon* and *12.4 Circular Planetary Orbits* for examples.)

Summary

For short displacements, the horizontal velocity of an object is constant and acts independently of the vertical force of gravity.

Likewise, an object projected at an angle to gravity can be broken into its horizontal and vertical components. The horizontal velocity component is also unaffected by the force of gravity.

However, for greater displacements, the curvature of the Earth comes into play, and since the angle of the force of gravity changes with displacement the horizontal velocity is no longer independent of gravity.

Mini-quiz to check your understanding

1. What happens when an object is projected perpendicular to gravity?

   a. It continually moves parallel to the ground

   b. Its horizontal velocity remains the same, but it falls according to gravity

   c. It slows down

2. Is the horizontal velocity constant when an object is projected at a downward angle?

   a. Yes, the horizontal component of the velocity remains constant

   b. It depends on the angle

   c. No, the horizontal velocity accelerates

3. If the Earth is a sphere, how can it be considered flat?

   a. Many people still believe the Earth is flat

   b. The Earth is not an exact sphere but is flat on one side

   c. Since the Earth is so large, its surface is approximately flat for short displacements

Answers

1b, 2a, 3c

Part 2: Derivations of Gravity Equations

It is worthwhile to see where equations come from, especially those used in gravity problems. Advanced Physics and Physical Science students are often required to derive equations to gain better understanding of the concepts involved.

The chapters in *Part 2* start with some fundamental assumptions and then go through the derivations of the gravity equations. Basic Calculus is used in some derivations.

Beginning students who have not yet taken Calculus can go over the material and still get an idea of how the equations evolved.

Part 2 Chapters

Chapters in Part 2 include:

2.1 Overview of Derivation of Gravity Equations

This chapter gives an overview of the results of the derivations of the gravity equations for velocity, displacement and time.

2.2 Derivation of Velocity-Time Equations

Starting with the equation $\mathbf{F} = \mathbf{mg}$, the this chapter shows the derivation for the velocity for a given time and then the equation for time for a given velocity.

2.3 Derivation of Displacement-Time Equations

This chapter starts from the result of the velocity-time derivation and continues to derive the equation of the displacement for a given time, the equation of the time for a given displacement and the equations concerning the effect of the initial velocity.

2.4 Derivation of Displacement-Velocity Equations

Starting with equations derived in the previous two chapters, the equation of the displacement for a given velocity, equation of the velocity for a given displacement and the equations concerning the effect of the initial velocity are derived in this chapter.

2.1 Overview of Gravity Equation Derivations

Starting with the assumption that the acceleration due to gravity is a constant value, you can derive equations that define the relationships between velocity, displacement and time for an object moving under the influence of gravity.

An initial velocity factor is included in these equations.

You use basic Calculus to determine the equations for the relationship between velocity and time for an object that is dropped, thrown downward or projected upward. From the velocity equation, the displacement-time relationship is derived.

Then the displacement-velocity equations are obtained from the previous two derivations.

Overview of velocity-time relationships

The velocity and time relationships as a result of the force of gravity are based on the fact that the acceleration due to gravity, **g**, is a constant value.

Since acceleration is the change in velocity with respect to time, the equation for the acceleration due to gravity is:

dv/dt = g

where

- **dv** is the first derivative or small change in velocity
- **dt** is the derivative or increment of time

Using Calculus, you integrate and derive the relationship between velocity and time for an object under the influence of gravity.

The results are:

$$v = gt + v_i$$

and

$$t = (v - v_i)/g$$

where

- v is the vertical velocity of the object in m/s or ft/s
- g is the acceleration due to gravity (9.8 m/s^2 or 32 ft/s^2)
- t is the time in seconds (s)
- v_i is the initial vertical velocity in m/s or ft/s

(See chapter *2.2 Derivation of Velocity-Time Equations* for details of the derivations.)

Overview of displacement-time relationships

The displacement a moving object travels in a given time is found by knowing that velocity is the change in displacement with respect to time:

$$v = dy/dt$$

Note: *Displacement* is a vector quantity denoting the change in position in a direction, whereas *displacement* is a scalar quantity that denotes the total change in position, independent of the path taken.

Substituting for v in the equation $v = gt + v_i$ and integrating, you get:

$$y = gt^2/2 + v_i t$$

Rearranging $y = gt^2/2 + v_i t$ and solving the quadratic equation for t gives you:

$$t = [-v_i \pm \sqrt{(v_i^2 + 2gy)}]/g$$

This equation can create some confusion because of the plus-or-minus sign.

If the object is thrown downward, the plus (+) sign is used. If the object is thrown upward, the sign depends on the object's position with respect to the starting point.

(See chapter *2.3 Derivation of Displacement-Time Equations* for details of the derivations.)

Overview of displacement-velocity relationships

To determine the displacement required to reach a given velocity, start with the equations $t = (v - v_i)/g$ and $y = gt^2/2 + v_i t$ from the previous derivations to get:

$$y = (v^2 - v_i^2)/2g$$

Solving for v, you get:

$$v = \pm\sqrt{(2gy + v_i^2)}$$

(See chapter *2.4 Derivation of Displacement-Velocity Equations* for details of the derivations.)

Summary

The relationships between the velocity of an object under the influence of gravity, the displacement and the time it takes to fall start with a basic equation $a = g$.

You use calculus to integrate equations, use algebra for substitutions and perform other operations to get the results.

Velocity with respect to time:

$$v = gt + v_i$$

Time with respect to velocity:

$$t = (v - v_i)/g$$

Displacement with respect to time:

$$y = gt^2/2 + v_i t$$

Time with respect to displacement:

$$t = [-v_i \pm \sqrt{(v_i^2 + 2gy)}]/g$$

Displacement with respect to velocity:

$$y = (v^2 - v_i^2)/2g$$

Velocity with respect to displacement:

$$v = \pm\sqrt{(2gy + v_i^2)}$$

Mini-quiz to check your understanding

1. What is the primary assumption in deriving the gravity equations?

    a. Calculus is similar to algebra, only harder

    b. Force is $\mathbf{F} = \mathbf{ma}$

    c. The acceleration due to gravity is a constant value

2. Why is displacement used in the equations instead of distance?

    a. Displacement is in a specific direction, while distance is independent of direction

    b. Either displacement or direction can be used

    c. Neither is ever used in the time equations

3. What is the displacement from the starting point when $\mathbf{v} = \mathbf{v}_i$?

    a. It is impossible for $\mathbf{v}$ to equal $\mathbf{v}_i$

    b. 0

    c. $2\mathbf{g}$

Answers

1c, 2a, 3b

2.2 Derivation of Velocity-Time Equations

The basis for the derivation of the gravity equation for the velocity an object reaches after a given time starts with the assumption that the acceleration due to gravity is a constant value.

Since acceleration is also the change in velocity for an increment of time, you use Calculus to integrate that change to get the velocity for a given elapsed time.

From the velocity equation, you can then determine the equation for the time it takes for the object to reach a given velocity from the starting point.

The derived equations are affected by the initial velocity of the object. This is important in later applications of the equations.

Basis for derivations

The derivations start with the assumption that the acceleration due to gravity **g** is a constant for displacements relatively close to Earth.

Acceleration is also the incremental change in velocity with respect to time:

$$\mathbf{a} = \mathbf{dv/dt}$$

where

- **a** is the acceleration
- **dv** is the first derivative of velocity **v** (a small change in velocity)
- **dt** is the first derivative of time **t** (a small time increment)

Since **g** is the acceleration due to gravity:

a = g

and

dv/dt = g

Multiply both sides of the equation by **dt** to get:

dv = g*dt

By using Calculus to integrate this equation, you can get the equations for velocity and time.

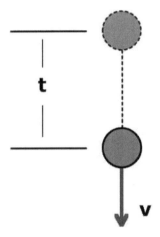

Velocity-time relationship

Derivation of velocity for a given time

To find the velocity for a given time you integrate both sides of the equation **dv = g*dt**

Integrate dv

First, integrate **dv** over the interval from **v = v$_i$** to **v = v.**

$\int$**dv = v − v$_i$**

where

- $\int$ is the integral sign, as used in Calculus

- **v** is the vertical velocity of the object
- **v**$_i$ is the initial vertical velocity of the object

Note: The *initial velocity* is the velocity at which the object is released after being accelerated from zero velocity. Initial velocity does not occur instantaneously.

Integrate g*dt

Then, integrate **g*dt** over the time interval from **t** = 0 to **t** = **t**.

∫**g*dt** = **gt** − 0

Resulting equation

The result of the two integrations is:

v − **v**$_i$ = **gt**

Thus, the general gravity equation for velocity with respect to time is:

v = **gt** + **v**$_i$

Derivation of time for a given velocity

To find the time it takes to reach a given velocity, you simply solve the equation **v** = **gt** + **v**$_i$ for **t**:

Subtract **v**$_i$ from both sides of the equation and then divide by **g**:

v − **v**$_i$ = **gt**

The general gravity equation for time with respect to velocity is:

t = (**v** − **v**$_i$)/**g**

Summary

Starting with the fact that the acceleration due to gravity **g** is considered a constant and knowing that acceleration is the change in velocity for a change in time, you can derive the gravity equations for the velocity with respect to time.

You can then determine the equation for the time to reach a given velocity.

The derived velocity-time equations are:

$$v = gt + v_i$$

$$t = (v - v_i)/g$$

Mini-quiz to check your understanding

1. What does **dv/dt** stand for?

   a. It is **d** times **v** divided by **d** times **t**

   b. It has no real meaning

   c. It is a small change in velocity divided by an increment in time

2. Why is **g*dt** integrated over the time interval 0 to **t**?

   a. You are starting from zero time to some arbitrary time

   b. The initial velocity is 0

   c. Because **g** can vary between 0 and **t**

3. How is the equation for time with respect to a given velocity determined?

   a. Integrate the velocity equation over time

   b. Take velocity for a given time equation and solve for t

   c. There is no equation for time with respect to a given velocity

Answers

1c, 2a, 3b

2.3 Derivation of Displacement-Time Equations

The basis for the derivation of the displacement-time gravity equations starts with the equation $\mathbf{v} = \mathbf{g}t + \mathbf{v}_i$ that was determined in the *Derivation of Velocity-Time Equations* chapter.

Since velocity is the change in displacement over an increment in time, you use Calculus to integrate that change and get the displacement for a given elapsed time. From the displacement equation, you can then determine the equation for the time it takes for the object to reach a given displacement from the starting point.

The derived equations are affected by the initial velocity of the object. This is important in later applications of the equations.

Basis for displacement-time derivations

To determine the displacement from the starting point for a given time, start with the equation:

$$\mathbf{v} = \mathbf{g}t + \mathbf{v}_i$$

(Obtained from chapter *2.2 Derivation of Velocity-Time Equations*)

where

- $\mathbf{v}$ is the vertical velocity in m/s or ft/s
- $\mathbf{g}$ is the acceleration due to gravity (9.8 m/s^2 or 32 ft/s^2)
- t is the time in seconds (s)
- $\mathbf{v}_i$ is the initial vertical velocity in m/s or ft/s

Velocity is also the incremental change in displacement with respect to time:

$$v = dy/dt$$

where

- **dy** is the first derivative of the vertical displacement, **y**
- **dt** is the first derivative of the elapsed time, **t**

By substituting combining these two equations and integrating, you can derive the displacement with respect to time. Then you can rearrange the equation and solve for **t** to get the time with respect to displacement.

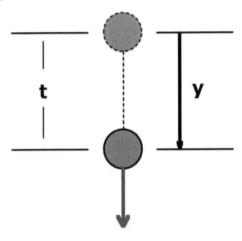

Displacement-time relationship

Derivation of displacement with respect to time

To obtain the displacement with respect to time, substitute for **v** in **v = gt + v$_i$**:

$$dy/dt = gt + v_i$$

Multiply both sides of the equation by **dt**:

$$dy = gt*dt + v_i*dt$$

Integrate **dy** over the interval from **y = 0** to **y = y**:

$$\int dy = y - 0$$

where

- $\int$ is the integral sign between the two limits
- **y** is the displacement from the starting point

Integrate **gt*dt** over the interval from **t = 0** to **t = t**:

$$\int gt^*dt = gt^2/2 - 0$$

Integrate $\mathbf{v_i^*dt}$ over the interval from **t = 0** to **t = t**:

$$\int v_i^*dt = v_i t - 0$$

The result of the integrations is the general gravity equation for the displacement with respect to time:

$$y = gt^2/2 + v_i t$$

Derivation of time with respect to displacement

You can find the time it takes for an object to travel a given displacement from the starting point by solving the following quadratic equation for **t**:

$$y = gt^2/2 + v_i t$$

Rearrange the equation by subtracting **y** from both sides of the equation and multiplying both sides by 2:

$$gt^2 + 2v_i t - 2y = 0$$

Solve the quadratic equation for **t**:

$$t = [-2v_i \pm \sqrt{(4v_i^2 + 8gy)}]/2g$$

Remove the square root of 4 from inside the square root or radical sign:

$$t = [-2v_i \pm 2\sqrt{(v_i^2 + 2gy)}]/2g$$

The resulting general gravity equation for time with respect to displacement is:

$$t = [-v_i \pm \sqrt{(v_i^2 + 2gy)}]/g$$

where

- $\pm$ means plus or minus
- $\sqrt{(v_i^2 + 2gy)}$ is the square root of the quantity $(v_i^2 + 2yg)$

The plus-or-minus sign means that in some situations, there can be two values for **t** for a given value of **y**.

Summary

The basis for the derivation of the displacement-time gravity equations starts with the equation $v = gt + v_i$. Since velocity is the change in displacement over an increment in time, you integrate that change and get the displacement for a given elapsed time.

From that displacement equation, you can then determine the equation for the time it takes for the object to reach a given displacement from the starting point.

The derived equations are:

$$y = gt^2/2 + v_i t$$

$$t = [-v_i \pm \sqrt{(v_i^2 + 2gy)}]/g$$

Mini-quiz to check your understanding

1. What is the starting point for deriving the gravity equation for displacement in a given time?

    a. Start by measuring the displacement

    b. Start with the equation $\mathbf{v} = \mathbf{g}t + \mathbf{v}_i$

    c. You must decide whether to throw the object up or down

2. Why must you do two integrations to find $\mathbf{y}$?

    a. It is a way of double-checking your answers

    b. Displacement always requires doing integration two times

    c. You integration both the left and right sides of the equal sign

3. How is the equation $\mathbf{g}t^2 + 2\mathbf{v}_i\mathbf{t} - 2\mathbf{y} = 0$ solved for $\mathbf{t}$?

    a. By using the formula for solving quadratic equations

    b. By trial and error

    c. You first solve for $\mathbf{g}$ and then substitute in values for $\mathbf{t}$

Answers

1b, 2c, 3a

2.4 Derivation of Displacement-Velocity Equations

The basis for the derivations of the displacement-velocity gravity equations starts with the equations $t = (v - v_i)/g$ and $y = gt^2/2 + v_i t$, as determined in the *Derivation of Velocity-Time Equations* and *Derivation of Displacement-Time Equations* chapters.

By substituting the time relationship in the displacement equation, you can determine the displacement with respect to velocity. From the derived displacement equation, you can then determine the equation for the velocity when the object reaches a given displacement from the starting point.

The derived equations are affected by the initial velocity of the object. This is important in later applications of the equations.

Basis for displacement-velocity derivations

To determine the displacement the object travels to reach a given velocity, start with the equations:

$$t = (v - v_i)/g$$

(Obtained from chapter *2.2 Derivation of Velocity-Time Gravity Equations*)

and

$$y = gt^2/2 + v_i t$$

(Obtained from chapter *2.3 Derivation of Displacement-Time Gravity Equations*)

where

- **t** is the time in seconds (s)
- **v** is the vertical velocity in m/s or ft/s
- v_i is the initial vertical velocity in m/s or ft/s
- **g** is the acceleration due to gravity (9.8 m/s^2 or 32 ft/s^2)
- **y** is the vertical displacement from the starting point

Note: The *initial velocity* is the velocity at which the object is released after being accelerated from zero velocity. Initial velocity does not occur instantaneously.

By substituting the equation for **t** in the equation for **y**, you can get the displacement with respect to velocity. Then by solving that equation for **v**, you get the velocity with respect to displacement equation.

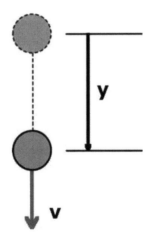

Displacement-velocity relationship

Derivation of displacement for a given velocity

To derive the displacement equation, you can start with the time equation:

$$t = (v - v_i)/g$$

Square both sides of the equation:

$$t^2 = (v - v_i)^2/g^2$$

$$t^2 = (v^2 - 2vv_i + v_i^2)/g^2$$

Consider the displacement equation:

$$y = gt^2/2 + v_i t$$

Substitute for t^2 and t from the above equations:

$$y = g\ (v^2 - 2vv_i + v_i^2)/2g^2 + v_i(v - v_i)/g$$

Multiply $v_i(v - v_i)/g$ by 2/2 and combine like terms:

$$y = (v^2 - 2vv_i + v_i^2)/2g + 2(vv_i - v_i^2)/2g$$

$$y = (v^2 - 2vv_i + v_i^2 + 2vv_i - 2v_i^2)/2g$$

The resulting general gravity equation for displacement with respect to velocity is:

$$y = (v^2 - v_i^2)/2g$$

Velocity for a given displacement

To get the velocity for a given displacement, multiply both sides of $y = (v^2 - v_i^2)/2g$ by $2g$ and solve for **v**:

$$2gy = (v^2 - v_i^2)$$

$$v^2 = 2gy + v_i^2$$

Take the square root of both sides of the equation to get the general gravity equation for velocity with respect to displacement:

$$v = \pm\sqrt{(2gy + v_i^2)}$$

where

- $\pm$ means plus or minus
- $\sqrt{(2gy + v_i^2)}$ is the square root of the quantity $(2gy + v_i^2)$

Summary

The gravity equation for the displacement an object travels from the starting point until it reaches a given velocity can be derived from the equations $t = (v - v_i)/g$ and $y = gt^2/2 + v_i t$.

This leads to the equation for the velocity when the object reaches a given displacement from the starting point.

The derived equations are:

$$y = (v^2 - v_i^2)/2g$$

$$v = \pm\sqrt{(2gy + v_i^2)}$$

Mini-quiz to check your understanding

1. What must be derived before finding the displacement-velocity relationship?

    a. Velocity-time and displacement-time relationships

    b. The exact value of acceleration due to gravity

    c. The initial velocity

2. What is the displacement equation if the initial velocity is zero?

    a. The equation becomes impossible to solve

    b. $y = v^2/2g$

    c. $y = -v_i^2/2g$

3. How can there be a positive and negative velocity?

    a. They mean the same thing

    b. It is an indication of before and after an object moves

    c. It is an indication of opposite directions of the velocity

Answers

1a, 2b, 3c

Part 3: Equations for Falling Objects

When you drop an object from some height, it falls toward the ground due to the force of gravity.

While *Part 2: Derivations of Gravity Equations*, provided the general equations for velocity, displacement and time, *Part 3* provides specific equations for falling objects.

The equations assume that the effect of air resistance is negligible.

Part 3 Chapters

Part 3 consists of the following chapters:

3.1 Overview of Gravity Equations for Falling Objects

This chapter gives a brief overview and states the velocity, displacement and time equations for falling objects.

3.2 Velocity Equations for Falling Objects

Given the time an object has fallen or the displacement is has traveled, this chapter gives the equation for its velocity, along with examples.

3.3 Displacement Equations for Falling Objects

This chapter provides the equations to determine the displacement a falling object travels after it reaches a given velocity or an elapsed time. Examples are included.

3.4 Time Equations for Falling Objects

Given the velocity an object reaches or the displacement it has fallen, this chapter provides the equations to determine the elapsed time of travel. Examples are included.

3.1 Overview of Equations for Falling Objects

A falling object is an object that you drop from some height above the ground. Since it is dropped, its initial velocity is zero ($v_i = 0$).

There are simple derived equations that allow you to calculate the velocity for a given time or displacement from the starting point, the displacement the object falls within a given time or when it has reached a given velocity, and the time it takes to reach a given velocity or displacement.

The overview of these equations is below.

Velocity equations

The equations for the velocity of a falling object are:

$$v = gt$$

$$v = \sqrt{(2gy)}$$

(See chapter *3.2 Velocity Equations for Falling Objects* for details on using these equations, as well as some examples.)

Displacement equations

The equations for the displacement that the object falls are:

$$y = gt^2/2$$

$$y = v^2/2g$$

(See chapter *3.3 Displacement Equations for Falling Objects* for details on using these equations, as well as some examples.)

Time equations

The equations for the elapsed time of a falling object are:

t = v/g

t = √(2y/g)

(See chapter *3.4 Time Equations for Falling Objects* for details on using these equations, as well as some examples.)

Summary

There are simple equations for falling objects that allow you to calculate the velocity and displacement traveled, as well as the time taken to achieve a given velocity or displacement.

Mini-quiz to check your understanding

1. If you know how long an object has been falling, what information can you calculate?

    a. The velocity and displacement traveled for that time period

    b. Which version of **g** to use

    c. The time it has been falling

2. How far has a falling object traveled when it reaches a velocity of 8 ft/s?

    a. 32 ft

    b. 8 ft

    c. 1 ft

3. What is the equation for the time it takes to fall a given displacement?

    a. $v = \sqrt{(2gy)}$

    b. $t = \sqrt{(2y/g)}$

    c. $t = v/g$

Answers

1a, 2c, 3b

3.2 Velocity Equations for Falling Objects

When you drop an object from some height above the ground, it has an initial velocity of zero. Simple equations allow you to calculate the velocity an object falls after a given period of time and the velocity it reaches at a given displacement. The equations assume that air resistance is negligible.

Examples demonstrate applications of the equations.

Velocity with respect to time

The general gravity equation for velocity with respect to time is:

$$v = gt + v_i$$

(See chapter *2.2 Derivation of Velocity-Time Equations* for details of the derivation.)

Since the initial velocity $v_i = 0$ for an object that is simply falling, the equation reduces to:

$$v = gt$$

where

- v is the vertical velocity of the object in meters/second (m/s) or feet/second (ft/s)
- g is the acceleration due to gravity (9.8 m/s^2 or 32 ft/s^2)
- t is the time in seconds (s) that the object has fallen

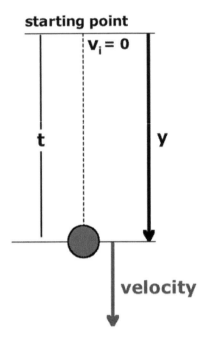

Velocity of falling object as function of time or displacement

Velocity with respect to displacement

The general gravity equation for velocity with respect to displacement is:

$$v = \pm\sqrt{(2gy + v_i^2)}$$

where

- $\pm$ means plus or minus
- $\sqrt{(2gy + v_i^2)}$ is the square root of the quantity $(2gy + v_i^2)$
- y is the vertical displacement in meters (m) or feet (ft)

(See chapter *2.4 Derivation of Displacement-Velocity Equations* for details of the derivation.)

Since $v_i = 0$, y is positive because it is below the starting point. Also, v is downward and positive. Only the $+$ term of $\pm$ applies.

Thus, the equation for the velocity of a falling object after it has traveled a certain displacement is:

v = √(2gy)

Examples

The following examples illustrate applications of the equations.

For a given time

What will be the velocity of an object after it falls for 3 seconds?

Solution

Substitute in the equation:

v = gt

If you use $\mathbf{g} = 9.8$ m/s^2, then $\mathbf{v} = (9.8$ m/s$^2)^*(3$ s$) = 29.4$ m/s.

If you use $\mathbf{g} = 32$ ft/s^2, then $\mathbf{v} = (32$ ft/s$^2)^*(3$ s$) = 96$ ft/s.

For a given displacement

What is the velocity of an object after it has fallen 100 feet?

Solution

Since **y** is in feet, $\mathbf{g} = 32$ ft/s^2. Substitute in the equation:

v = √(2gy)

$\mathbf{v} = \sqrt{[2^*(32 \text{ ft/s}^2)^*(100 \text{ ft})]}$

$\mathbf{v} = \sqrt{(6400 \text{ ft}^2/\text{s}^2)}$

v = 80 ft/s

Summary

There are simple equations for falling objects that allow you to calculate the velocity the object reaches for a given displacement or time. The equations are:

$$v = gt$$

$$v = \sqrt{(2gy)}$$

Mini-quiz to check your understanding

1. What velocity in m/s does an object reach in 10 seconds?

    a. 0.98 meters/second

    b. 9.8 meters per second

    c. 98 meters per second

2. What is the velocity of an object after it has fallen 400 feet?

    a. 160 ft/s

    b. 80 ft/s

    c. 400 ft/

3. If an object falls for 3 s, which velocity is greater: 29.4 m/s or 96 ft/s?

    a. 96 ft/s is greater, because 96 is more than 29.4

    b. 29.4 is greater, because meters are larger than feet

    c. They are the same velocity, except with different units of measurement

Answers

1c, 2a, 3c

3.3 Displacement Equations for Falling Objects

The displacement of an object is the change in position from the starting point in a specific direction and can be represented as a vector. It is different from distance, where direction is not indicated.

When you drop an object from some height above the ground, it has an initial velocity of zero. Simple equations allow you to calculate the displacement the object falls until it reaches a given velocity or after a given period of time. The equations assume that air resistance is negligible.

Examples demonstrate applications of the equations.

Displacement with respect to velocity

The general gravity equation for displacement with respect to velocity is:

$$\mathbf{y} = (\mathbf{v}^2 - \mathbf{v_i}^2)/2\mathbf{g}$$

(See chapter *2.4 Derivation of Displacement-Velocity Equations* for details of the derivation.)

Since the initial velocity $\mathbf{v_i} = 0$ for a dropped object, the equation reduces to:

$$\mathbf{y} = \mathbf{v}^2/2\mathbf{g}$$

where

- $\mathbf{y}$ is the vertical displacement in meters or feet
- $\mathbf{v}$ is the vertical velocity in m/s or ft/s
- $\mathbf{g}$ is the acceleration due to gravity (9.8 m/s^2 or 32 ft/s^2)

Since the object is moving downward from the starting point, both **y** and **v** are positive numbers.

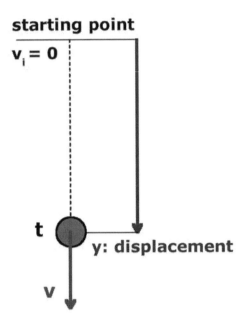

Displacement of falling object as function of velocity or time

Displacement with respect to time

The general gravity equation for the displacement with respect to time is:

$$y = gt^2/2 + v_i t$$

(See chapter *2.3 Derivation of Displacement-Time Equations* for details of the derivation.)

Since $v_i = 0$ for a dropped object, the equation reduces to:

$$y = gt^2/2$$

where **t** is the time in seconds (s).

Examples

The following examples illustrate applications of the equations.

Given the velocity

If **v** = 75 ft/s, how far has the object fallen?

Solution

Since **v** is in ft/s, **g** = 32 ft/s². Substitute values in the equation:

$$y = v^2/2g$$

$$y = (75 \text{ ft/s})*(75 \text{ ft/s})/2*(32 \text{ ft/s}^2)$$

$$y = (5625 \text{ ft}^2/\text{s}^2)/(64 \text{ ft/s}^2)$$

$$y = 87.89 \text{ ft}$$

Given the elapsed time

If **t** = 4 s and **g** = 9.8 m/s², find how far the object has fallen.

Solution

Substitute values in the equation to obtain the displacement:

$$y = gt^2/2$$

$$y = (9.8 \text{ m/s}^2)*(16 \text{ s}^2)/2 = 78.4 \text{ m}$$

Summary

Displacement is the change in position from the starting point in a specific direction. The following equations allow you to calculate the displacement the object falls until it reaches a given velocity or after a given period of time:

$$y = v^2/2g$$

$$y = gt^2/2$$

Mini-quiz to check your understanding

1. If an object is falling at a rate of 9.8 m/s, how far has it fallen?

   a. 19.6 meters

   b. 9.8 meters

   c. 4.9 meters

2. If an object falls for 3 seconds, how many feet has it fallen?

   a. 144 feet

   b. 144 meters

   c. 16 feet

3. If the velocity is 50 miles per hour (mph), what do you do to find the displacement the object has fallen?

   a. You can't solve the problem

   b. Convert mph to ft/s

   c. Just use 50 mph in the equation

Answers

1c, 2a, 3b

3.4 Time Equations for Falling Objects

When you drop an object from some height above the ground, it has an initial velocity of zero. Simple equations allow you to calculate the time it takes for a falling object to reach a given velocity and the time it takes to reach a given displacement. The equations assume that air resistance is negligible.

Examples demonstrate applications of the equations.

Time with respect to velocity

The general gravity equation for elapsed time with respect to velocity is:

$$t = (v - v_i)/g$$

(See chapter *2.2 Derivation of Velocity-Time Equations* for details of the derivation.)

Since the initial velocity $v_i = 0$ for an object that is simply falling, the equation reduces to:

$$t = v/g$$

where

- t is the time in seconds
- v is the vertical velocity in meters/second (m/s) or feet/second (ft/s)
- g is the acceleration due to gravity (9.8 m/s^2 or 32 ft/s^2)

Since the object is moving in the direction of gravity, v is a positive number.

Time with respect to displacement

The general gravity equation for the elapsed time with respect to displacement is:

$$t = [\, -v_i \pm \sqrt{(v_i^2 + 2gy)}\,]/g$$

where

- $\pm$ means plus-or-minus
- $\sqrt{(v_i^2 + 2gy)}$ is the square root of the quantity $(v_i^2 + 2gy)$
- y is the vertical displacement in meters (m) or feet (ft)

(See *2.3 Derivation of Displacement-Time Gravity Equations* for details of the derivation.)

When the object is simply dropped, the initial velocity is zero ($v_i = 0$) and the equation for elapsed time becomes:

$$t = \pm \sqrt{(2gy)}/g$$

Since time t is always positive, the equation is:

$$t = \sqrt{(2gy)}/g$$

Change g to $\sqrt{(g^2)}$ and simplify the equation:

$$t = \sqrt{(2gy)}/\sqrt{(g^2)}$$

Thus, the resulting time equation is:

$$t = \sqrt{(2y/g)}$$

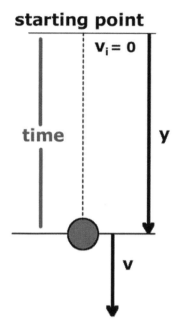

starting point

$v_i = 0$

time

y

v

Elapsed time as a function of velocity or displacement

Examples

The following are examples to illustrate application of the equations.

Given the velocity

How long does it take for a falling object to reach 224 ft/s?

Solution

Since **v** is in ft/s, **g** = 32 ft/s². Substitute values in the equation:

$$t = v/g$$

$$t = (224 \text{ ft/s})/(32 \text{ ft/s}^2)$$

$$t = 7 \text{ s}$$

Given the displacement

How long does it take for an object to fall 200 meters?

Solution

Since displacement is in meters, **g** = 9.8 m/s². Substitute values in the equation:

$$t = \sqrt{(2y/g)}$$

$$t = \sqrt{[2*(200\ m)/(9.8\ m/s^2)]}$$

$$t = \sqrt{(40.8\ s^2)}$$

$$t = 6.39\ s$$

Summary

There are simple equations for falling objects that allow you to calculate the time taken to achieve a given velocity or displacement. These equations are:

$$t = v/g$$

$$t = \sqrt{(2y/g)}$$

Mini-quiz to check your understanding

1. How long does it take a falling object to reach 196 m/s?

    a. 2 minutes

    b. 200 seconds

    c. 20 seconds

2. How long does it take a falling object to reach 19.6 m?

    a. 2 seconds

    b. 4 seconds

    c. 8 seconds

3. How long does it take a falling object to reach 196 ft/s?

    a. 20 seconds

    b. 6.125 seconds

    c. 3.98 seconds

Answers

1c, 2a, 3b

Part 4: Equations for Objects Projected Downward

While *Part 2: Derivations of Gravity Equations*, provided the general equations for velocity, displacement and time, *Part 4* includes specific equations for objects that are initially projected downward.

The equations assume that the effect of air resistance is negligible.

Part 4 Chapters

Part 4 consists of the following chapters:

4.1 Overview of Equations for Objects Projected Downward

This chapter gives a brief overview and states the velocity, displacement and time equations for objects projected downward.

4.2 Velocity Equations for Objects Projected Downward

The velocity of an object projected downward at some initial velocity can be determined from the displacement it has fallen or the elapsed time of its fall. There are also examples of applications of the equations.

4.3 Displacement Equations for Objects Projected Downward

This chapter provides the equations to determine the displacement an object projected downward travels after it reaches a given velocity or an elapsed time. Examples are included.

4.4 Time Equations for Objects Projected Downward

The time it takes an object projected downward to reach a given displacement or velocity is determined by its initial velocity. This chapter provides the equations to determine the elapsed time of travel. Examples are included.

4.1 Overview of Equations for Objects Projected Downward

When you throw or project an object downward, the object typically starts at zero velocity and is accelerated until it is released at some initial velocity.

The object then accelerates due to the force of gravity, with the initial velocity adding to the result of the acceleration.

There are simple equations that allow you to calculate the velocity and displacement traveled, as well as the time taken to achieve a given velocity or displacement.

This chapter is an overview of the equations and has references to the other chapters, which provide the details.

Velocity equations

The equations for the velocity of an object projected downward at an initial vertical velocity of v_i are:

$$v = gt + v_i$$

$$v = \sqrt{(2yg + v_i{}^2)}$$

(See chapter *4.2 Velocity Equations for Objects Projected Downward* for details on using these equations, as well as some examples.)

Displacement equations

The equations for the vertical displacement traveled of an object projected downward at an initial velocity of v_i are:

$$y = gt^2/2 + v_i t$$

$$y = (v^2 - v_i{}^2)/2g$$

(See chapter *4.3 Displacement Equations for Objects Projected Downward* for details on using these equations, as well as some examples.)

Time equations

The equations for the elapsed time of an object projected downward at an initial velocity of v_i are:

$$t = (v - v_i)/g$$

$$t = [-v_i + \sqrt{(v_i^2 + 2gy)}]/g$$

(See chapter *4.4 Time Equations for Objects Projected Downward* for details on using these equations, as well as some examples.)

Summary

There are equations for objects projected downward that allow you to calculate the velocity and displacement traveled, as well as the time taken to achieve a given velocity or displacement.

Mini-quiz to check your understanding

1. If the initial velocity is 32 ft/s, what is the velocity at 1 second?

    a. 64 ft/s

    b. 32 ft/s

    c. You also need the displacement to solve the problem

2. If the initial velocity is 10 m/s, what is the displacement at $v = 20$ m/s?

    a. $y = (20 - 10)/19.6$ meters

    b. $y = (400 - 100)/19.6$ meters

    c. $y = (400 - 100)/64$ feet

3. How long does it take a ball thrown downward to hit the ground?

    a. It only depends on the displacement to the ground

    b. 12.3 seconds

    c. It depends on the initial velocity and the height from which it was thrown

Answers

1a, 2b, 3c

4.2 Velocity Equations for Objects Projected Downward

When you throw or project an object downward, it is accelerated until it is released at some initial velocity.

If you know this initial velocity, there are simple derived equations that allow you to calculate the velocity when the object reaches a given displacement from the starting point or when it reaches a given elapsed time. Examples illustrate these equations.

Velocity with respect to displacement

The general gravity equation for velocity with respect to displacement is:

$$v = \pm\sqrt{(2gy + v_i^2)}$$

where

- $\pm$ means plus or minus
- v is the vertical velocity in m/s or ft/s
- $\sqrt{(2gy + v_i^2)}$ is the square root of the quantity $(2gy + v_i^2)$
- y is the vertical displacement in m or ft
- g is the acceleration due to gravity (9.8 m/s^2 or 32 ft/s^2)
- v_i is the initial vertical velocity of the object

(See chapter *2.4 Derivation of Displacement-Velocity Equations* for details of the derivation.)

Since v is a downward vector, it has a positive value. Likewise, y and v_i are positive numbers. Thus, only the + version of the equation applies:

$$v = \sqrt{(2gy + v_i^2)}$$

Velocity with respect to time

The general gravity equation for velocity with respect to time is:

$$v = gt + v_i$$

where **t** is the time the object has traveled in seconds (s).

(See chapter *2.2 Derivation of Velocity-Time Equations* for details of the derivation.)

This same equation applies for an object projected downward.

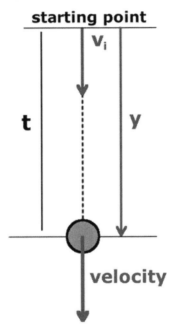

Velocity as a function of displacement or time

Examples

The following examples illustrate applications of the equations.

For a given displacement

Find the velocity of a rock that is thrown down at 2 m/s after it has traveled 2 meters.

Solution

You are given that $v_i = 2$ m/s and $y = 2$ m. Since v_i is in m/s and y is in meters, then $g = 9.8$ m/s^2. The equation to use is:

$$v = \sqrt{(2gy + v_i{}^2)}$$

Substitute values in the equation:

$$v = \sqrt{[2*(9.8 \text{ m/s}^2)*(2 \text{ m}) + (2 \text{ m/s})^2]}$$

$$v = \sqrt{(39.2 \text{ m}^2/\text{s}^2 + 4 \text{ m}^2/\text{s}^2)}$$

$$v = \sqrt{(43.2 \text{ m}^2/\text{s}^2)}$$

Thus, the velocity is:

$$v = 6.57 \text{ m/s}$$

For a given time

Suppose you throw the object downward at 10 m/s. Find its velocity after 4 seconds.

Solution

You are given that $v_i = 10$ m/s and $t = 4$ s. Since v_i is in m/s, then $g = 9.8$ m/s^2. The equation to use is:

$$v = gt + v_i$$

Substitute values in the equation:

$$v = (9.8 \text{ m/s}^2)*(4 \text{ s}) + 10 \text{ m/s}$$

$$v = 39.2 \text{ m/s} + 10 \text{ m/s}$$

The velocity after 4 seconds is:

$$v = 49.2 \text{ m/s}$$

Summary

You can calculate the velocity when an object that is projected downward reaches a given displacement from the starting point or when it reaches a given elapsed time from the equations:

$$v = \sqrt{(2gy + v_i^2)}$$

$$v = gt + v_i$$

Mini-quiz to check your understanding

1. What is the velocity of an object thrown downward at 10 ft/s after it falls 5 feet?

    a. $v = \sqrt{(2.9.8*5 + 25)}$ m/s

    b. $v = (32*5 + 10)$ ft/s

    c. $v = \sqrt{(2*32*5 + 25)}$ ft/s

2. If the initial velocity is 30 m/s, what is the velocity of an object that has been falling for 10 s?

    a. 128 meters/second

    b. 304 m/s

    c. 100 meters per second

3. If $v_i = 2$ m/s, what is the velocity after it has traveled $y = 4$ meters?

    a. $v = \sqrt{(2*32*4 + 4)}$ m/s

    b. $v = \sqrt{(2*9.8*4 + 4)}$ m/s

    c. $v = \sqrt{(2*9.8*4 + 2)}$ m/s

Answers

1c, 2a, 3b

4.3 Displacement Equations for Objects Projected Downward

The displacement of an object is the change in position from the starting point in a specific direction and can be represented as a vector. It is different from distance, where direction is not indicated.

When you throw or project an object downward, it is accelerated until it is released at some initial velocity. If you know this initial velocity, there are simple derived equations that allow you to calculate the displacement traveled from the starting point when the object reaches a given velocity or when it reaches a given elapsed time.

Examples illustrate these equations.

Displacement with respect to velocity

The general gravity equation applies in the case where you project the object downward and release it at an initial velocity v_i. The result is that v_i is a positive number, as are y and v:

$$y = (v^2 - v_i^2)/2g$$

where

- y is the vertical displacement from the starting point in meters (m) or feet (ft)
- v is the vertical velocity in meters/second (m/s) or feet/second (ft/s)
- v_i is the initial vertical velocity in m/s or ft/s
- g is the acceleration due to gravity (9.8 m/s^2 or 32 ft/s^2)

(See chapter *2.4 Derivation of Displacement-Velocity Equations* for details of the derivation.)

Displacement with respect to time

If you project an object downward, the initial velocity is a positive number. The equation for the displacement traveled within a given time is:

$$y = gt^2/2 + v_i t$$

where **t** is the time the object has fallen in seconds (s).

(See chapter *2.3 Derivation of Displacement-Time Equations* for details of the derivation.)

Since the displacement is below the starting point, **y** is a positive number.

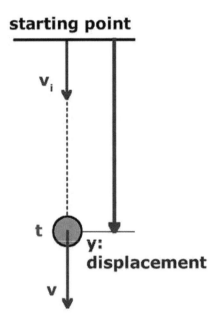

Downward displacement as a function of velocity and time

Examples

The following examples illustrate applications of the equations.

Displacement for a given velocity

If you throw an object downward at 10 m/s, find the minimum elevation from which you must throw the object so that it reaches 50 m/s.

Solution

You are given that v_i = +10 m/s and v = 50 m/s. Since v_i and v are in m/s, g = 9.8 m/s². The equation to use is:

$$y = (v^2 - v_i^2)/2g$$

Substitute values in the equation:

$$y = [(50 \text{ m/s})^2 - (10 \text{ m/s})^2]/2*(9.8 \text{ m/s}^2)$$

$$y = [(2500 \text{ m}^2/\text{s}^2) - (100 \text{ m}^2/\text{s}^2)]/(19.6 \text{ m/s}^2)$$

$$y = (2400 \text{ m}^2/\text{s}^2)/(19.6 \text{ m/s}^2)$$

The elevation is:

$$y = 122.4 \text{ m}$$

Displacement for a given time

If you throw an object downward at 30 ft/s and it travels for 4 seconds, find the displacement.

Solution

You are given that v_i = 30 ft/s and t = 4 s. Since v_i is in ft/s, g = 32 ft/s². The equation to use is:

$$y = gt^2/2 + v_i t$$

Substitute values in the equation:

$$y = [(32 \text{ ft/s}^2)*(4 \text{ s})^2]/2 + (30 \text{ ft/s})*(4 \text{ s})$$

$$\mathbf{y} = (32 \text{ ft/s}^2)*(16 \text{ s}^2)/2 + 120 \text{ ft}$$

$$\mathbf{y} = (512 \text{ ft})/2 + 120 \text{ ft}$$

The displacement is:

$$\mathbf{y} = 376 \text{ ft}$$

Summary

You can calculate the displacement from the starting point when an object that is projected downward reaches a given velocity or when it reaches a given elapsed time from the equations:

$$\mathbf{y} = (\mathbf{v}^2 - \mathbf{v}_i{}^2)/2\mathbf{g}$$

$$\mathbf{y} = \mathbf{g}t^2/2 + \mathbf{v}_i t$$

Mini-quiz to check your understanding

1. If you throw an object downward at 8 ft/s, how far does it travel when it reaches 20 ft/s?

    a.   20 feet

    b.   5.25 feet

    c.   2 feet

2. If you throw an object downward at 10 m/s, how far does it travel in 2 s?

    a.   39.6 meters

    b.   10 meters

    c.   1 meter

3. If you throw an object downward at 10 ft/s, how far does it travel in 2 s?

    a.   8.4 feet

    b.   84 feet

    c.   840 feet

Answers

1b, 2a, 3b

4.4 Time Equations for Objects Projected Downward

When you throw or project an object downward, it is accelerated until it is released at some initial velocity. If you know this initial velocity, there are simple derived equations that allow you to calculate the time it takes for it to reach a given velocity or when it reaches a given displacement from the starting point.

Examples illustrate these equations

Time with respect to velocity

The equation for the time it takes an object thrown downward at velocity v_i to reach velocity v is:

$$t = (v - v_i)/g$$

where

- **t** is the time in seconds (s)
- **v** is the vertical velocity of the falling object in feet/second (ft/s) or meters/second (m/s)
- v_i is the initial vertical velocity the object has been projected downward in ft/s or m/s
- **g** is the acceleration due to gravity; (32 ft/s^2 or 9.8 m/s^2)

(See chapter *2.2 Derivation of Velocity-Time Equations* for details of the derivation.)

Since the object is moving in the direction of gravity, **v** and v_i are positive numbers.

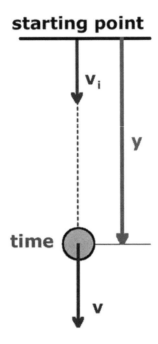

Time as a function of velocity and displacement

Time with respect to displacement

The general gravity equation for the time with respect to displacement is:

$$t = [-v_i \pm \sqrt{(v_i^2 + 2gy)}]/g$$

where

- $\pm$ means plus or minus
- **y** is the vertical displacement in feet or meters
- $\sqrt{(v_i^2 + 2gy)}$ is the square root of the quantity $(v_i^2 + 2gy)$

(See chapter *2.3 Derivation of Displacement-Time Equations* for details of the derivation.)

Since v_i is downward, it has a positive value and $-v_i$ is obviously negative. This means that the + version of the equation must be used in order to make t a positive number. The equation is then:

$$t = [-v_i + \sqrt{(v_i^2 + 2gy)}]/g$$

Since **y** is below the starting point, it also is a positive number.

Examples

The following examples illustrate application of the equations.

Time for a given velocity

If you throw a ball downward from a tall building at 5 ft/s, find the time it takes for the ball to reach a velocity of 101 ft/s.

Solution

You are given that $v_i = +5$ ft/s and $v = 101$ ft/s. Since v_i and v are in ft/s, then $g = 32$ ft/s^2. The equation to use is:

$$t = (v - v_i)/g$$

Substitute values in the equation:

$$t = (101 \text{ ft/s} - 5 \text{ ft/s})/(32 \text{ ft/s}^2)$$

$$t = (96 \text{ ft/s})/(32 \text{ ft/s}^2)$$

$$t = 3 \text{ s}$$

Time for a given displacement

If you throw an object downward from a high building at 5 m/s, find the time it takes to fall 50 m.

Solution

You are given that $v_i = +5$ m/s and $y = 50$ m. Since v_i in m/s and **y** is in meters, then $g = 9.8$ m/s^2. The equation to use is:

$$t = [-v_i + \sqrt{(v_i^2 + 2gy)}]/g$$

Substitute values in the equation:

$$t = [-5 \text{ m/s} + \sqrt{\{(25 \text{ m/s})^2 + 2*(9.8 \text{ m/s}^2)*(50 \text{ m})\}}]$$
$$/(9.8 \text{ m/s}^2)$$

$t = [-5 \text{ m/s} + \sqrt{(625 \text{ m}^2/\text{s}^2 + 980 \text{ m}^2/\text{s}^2)}]/(9.8 \text{ m/s}^2)$

$t = [-5 \text{ m/s} + \sqrt{(1605 \text{ m}^2/\text{s}^2)}]/(9.8 \text{ m/s}^2)$

$t = [-5 \text{ m/s} + 40.1 \text{ m/s}]/(9.8 \text{ m/s}^2)$

$t = (35.1 \text{ m/s})/(9.8 \text{ m/s}^2)$

$t = 3.58 \text{ s}$

(Whew!)

Summary

You can calculate the time it takes an object that is projected downward to reach a given velocity or reach a given displacement from the starting point from the equations:

$t = (v - v_i)/g$

$t = [-v_i + \sqrt{(v_i^2 + 2gy)}]/g$

Mini-quiz to check your understanding

1. How long does it take to reach 118 m/s if the initial velocity downward is 20 m/s?

   a. 1 second

   b. 9.8 seconds

   c. 10 seconds

2. If $t = [-v_i + \sqrt{(v_i^2 + 2gy)}]/g$, how long does it take to reach 64 ft if the initial velocity is +32 ft/s?

   a. $t = [-32 + \sqrt{(32^2 + 2*32*64)}]/32$ seconds

   b. $t = [-32 + \sqrt{(32^2 + 2*9.8*64)}]/9.8$ seconds

   c. $t = [32 + \sqrt{(32^2 + 2*32*64)}]/32$ seconds

3. What is the time when the displacement is zero?

   a. $t = [-v_i + \sqrt{(v_i^2)}]/g = 2v_i/g$ seconds

   b. $t = [-v_i + \sqrt{(v_i^2)}]/g = 0$ seconds

   c. **y** cannot equal zero

Answers

1c, 2a, 3b

Part 5: Equations for Objects Projected Upward

When you project an object upward, it is accelerated until it is released at some initial velocity.

Since it is moving in the opposite direction of the force of gravity, it has an initial velocity less than zero. The object slows until it reaches a maximum displacement, after which it falls toward the ground due to the force of gravity.

Part 2: Derivations of Gravity Equations, provided the general equations for velocity, displacement and time. This part of the book includes specific equations for objects that are initially projected upward, as well as examples of applying those equations.

Assume that the effect of air resistance is negligible and that the object is projected upward to not much more than 40 miles or 64 kilometers above the ground.

Part 5 Chapters

Part 5 consists of the following chapters:

5.1 Overview of Equations for Objects Projected Upward

This chapter gives a brief overview and states the velocity, displacement and time equations for objects projected upward.

5.2 Velocity Equations for Objects Projected Upward

The velocity of an object projected downward at some initial velocity can be determined by the elapsed time of its travel, by the displacement it traveled above the starting point and the displacement it traveled below the starting point. Examples are included.

5.3 Displacement Equations for Objects Projected Upward

This chapter provides the equations to determine the displacement from the starting point that an object projected upward has traveled after it reaches a given velocity or an elapsed time. It also shows how to calculate the total displacement traveled. Examples are included.

5.4 Time Equations for Objects Projected Upward

Given the initial velocity an object that has been projected upward, this chapter provides equations for the time it takes to reach a given velocity, as well as the time for displacements above and below the starting point. Examples are included.

5.1 Overview of Equations for Objects Projected Upward

When an object is projected upward and released at some initial velocity, it is moving in the opposite direction of the force of gravity.

Since our convention states that the direction of gravity is positive, the upward initial velocity is then a negative number.

> (See chapter *1.3 Convention for Direction in Gravity Equations* for more information.)

The object moves upward, slowing down from its initial velocity until it reaches its peak or maximum displacement. Then it falls back down to the ground.

Knowing the initial velocity, you can calculate the velocity, displacement and time of the object during its flight.

It is assumed that the air resistance on the object is negligible.

Also, there is a restriction on how high the object can be projected. At heights above 64 km or 40 mi, the value of the acceleration due to gravity changes enough to make your calculations inaccurate.

> (See chapter *1.4 Gravity Constant* for more information.)

This chapter is an overview of the equations and has references to the other chapters, which provide the details.

Velocity equations

The equations for the velocity of an object projected upward at an initial vertical velocity of v_i are:

With respect to time

$$v = gt + v_i$$

With respect to time

$$v = -\sqrt{(2gy + v_i^2)} \text{ (going up)}$$

$$v_m = 0 \text{ (at maximum displacement)}$$

$$v = +\sqrt{(2gy + v_i^2)} \text{ (coming down)}$$

(See chapter *5.2 Velocity Equations for Objects Projected Upward* for details.)

Displacement equations

The equations for the displacement from the starting point of an object projected upward at an initial velocity of v_i are:

With respect to displacement

$$y = (v^2 - v_i^2)/2g$$

$$y_m = -v_i^2/2g \text{ (peak or maximum displacement)}$$

With respect to time

$$y = gt^2/2 + v_i t$$

$$y_m = -gt_m^2/2 \text{ (maximum displacement)}$$

Total distance traveled

$$d_u = |y| \text{ (going upward to maximum displacement)}$$

$$d = |2y_m| + y \text{ (sum of going up and coming down)}$$

(See chapter *5.3 Displacement Equations for Objects Projected Upward* for details.)

Time equations

The equations for the time an object projected upward at an initial velocity of v_i travels are:

With respect to velocity

$t = (v - v_i)/g$

$t_m = -v_i/g$ (At maximum displacement)

With respect to displacement

$t = [-v_i - \sqrt{(v_i^2 + 2gy)}]/g$ (going up)

$t_m = \sqrt{(-2y_m/g)}$ (At maximum displacement)

$t = [-v_i + \sqrt{(v_i^2 + 2gy)}]/g$ (coming down)

(See chapter *5.4 Time Equations for Objects Projected Upward* for details.)

Summary

When an object is projected upward and released at some initial velocity, it is moving in the opposite direction of the force of gravity, and the initial velocity is negative.

The object moves upward, reaches its peak or maximum displacement and then falls toward the ground. Knowing the initial velocity, you can calculate the velocity, displacement and time of the object during its flight.

Mini-quiz to check your understanding

1. For an object projected upward, when $t = 0$, what does v equal?

    a. $v = v_i$

    b. $v = 0$

    c. $v = -v_i$

2. What is the equation for the maximum height of a ball thrown upward at 10 m/s?

    a. $y_m = (v^2 - v_i^2)/2g$

    b. $y_m = -v_i^2/2g$

    c. $y_m = gt^2/2 + v_i t$

3. If an object is thrown upward at -40 ft/s, what is the elapsed time when $v = -8$ ft/s?

    a. $t = (-8 - 40)/32 = -1.5$ seconds

    b. $t = (8 + 40)/32 = 1.5$ seconds

    c. $t = (-8 + 40)/32 = 1$ second

Answers

1a, 2b, 3c

5.2 Velocity Equations for Objects Projected Upward

When you project an object upward and release it at its initial velocity, it is moving in the opposite direction of the force of gravity.

Thus the initial velocity is negative. The velocity of the object is also negative on the way up but positive on the way down.

The object slows down as it moves upward until it reaches a maximum height, at which time the velocity is zero. Then the velocity increases as the object falls toward the ground.

Derived equations allow you to calculate the velocity of an object projected upward with respect to time, as well as the velocity at displacements both above and below the starting point.

Velocity with respect to time

The general gravity equation for the velocity with respect to time of an object thrown upward and released at an initial velocity is:

$$v = gt + v_i$$

where

- v is the vertical velocity in meters/second (m/s) or feet/second (ft/s)
- g is the acceleration due to gravity (9.8 m/s² or 32 ft/s²)
- t is the time in seconds (s)
- v_i is the upward initial vertical velocity in m/s or ft/s

(See chapter *2.2 Derivation of Velocity-Time Equations* for details of the derivation.)

When you project the object upward, the initial velocity when you release it is negative or less than zero ($v_i < 0$). The resulting velocities will be negative ($v < 0$) when the object is moving upward, zero ($v = 0$) at the maximum displacement or positive ($v > 0$) when the object is moving downward, depending on the value for elapsed time.

Time to maximum displacement

When determining the velocity at various times, it is convenient to know the time required to reach the maximum displacement, when the velocity is $v = 0$:

$$gt_m + v_i = 0$$

$$gt_m = -v_i$$

$$t_m = -v_i/g$$

where t_m is the time to the maximum displacement.

Note: Since v_i is a negative number, $-v_i$ is a positive number

Example

If the initial velocity of an object is in the upward direction at 19.6 m/s, what is the velocity at various times?

Solution

$v_i = -19.6$ m/s and $g = 9.8$ m/s^2. Substitute for v_i and g in the equation in order to get a formula in terms of t:

$$v = gt + v_i$$

$$v = (9.8 \text{ m/s}^2)(t \text{ s}) + (-19.6 \text{ m/s})$$

Simplify:

$$v = (9.8t - 19.6) \text{ m/s}$$

The table below shows the velocites for different values of t:

Table

t = 0 s	v = −19.6 m/s	Moving upward from starting point
t = 1 s	v = −9.8 m/s	Object moving upward
t_m = 2 s	v = 0 m/s	At peak or maximum height
t = 3 s	v = 9.8 m/s	Object moving downward
t = 4 s	v = 19.6 m/s	Passing starting point
t = 6 s	v = 39.2 m/s	Continuing downward

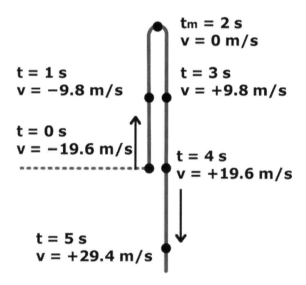

Velocities of object projected upward at different times

Velocity for displacements moving upward

The general gravity equation for the velocity of an object with respect to the displacement—or movement from the starting point—is:

$$v = \pm\sqrt{(2gy + v_i^2)}$$

where

- $\pm$ means plus or minus
- $\sqrt{(2yg + v_i^2)}$ is the square root of the quantity $(2yg + v_i^2)$
- y is the vertical displacement in m or ft

(See chapter *2.4 Derivation of Displacement-Velocity Equations* for details of the derivation.)

Since the velocity is negative ($v < 0$) on the way up, the negative ($-$) version of the equation is used:

$$v = -\sqrt{(2gy + v_i^2)}$$

Also, note that the displacement is negative ($y < 0$) above the starting point.

Maximum displacement with respect to velocity

At the maximum displacement, the velocity is $v = 0$. Thus:

$$-\sqrt{(2gy_m + v_i^2)} = 0$$

$$2gy_m + v_i^2 = 0$$

$$2gy_m = -v_i^2$$

$$y_m = -v_i^2/2g$$

where y_m is the maximum displacement.

If you substitute values for the displacement where $y < y_m$, the quantity $(2gy + v_i^2)$ becomes negative, resulting in the value of $\sqrt{(2gy + v_i^2)}$ being imaginary or impossible.

Example

If $v_i = -64$ ft/s, find the values of v for various displacements y moving upward.

Solution

Since $g = 32$ ft/s^2, substitute for **g** and v_i in the equation to get a formula in terms of **y**:

$$v = -\sqrt{(2gy + v_i^2)}$$

$$v = -\sqrt{[2*(32\ ft/s^2)(y\ m) + (-64\ ft/s)^2]}$$

$$v = -\sqrt{(64y\ ft^2/s^2 + 4096\ ft^2/s^2)}$$

Since $\sqrt{(ft^2/s^2)}$ = ft/s, you get:

$$v = -\sqrt{(64y + 4096)}\ ft/s$$

Substitute values for **y** in the formula, remembering that **y** is negative above the starting point:

y = 0 ft	**v** = −64 ft/s	Moving upward from starting point
y = −32 ft	**v** = −45.3 ft/s	Object moving upward
y = −64 ft	**v** = 0 ft/s	At peak or maximum displacement
y = −80 ft	--	$v = -\sqrt{(-1024)}$ - impossible

Velocity for displacements moving downward

Below the maximum displacement, **v** is positive, since the object is moving in the direction of gravity. This means that the positive (+) version of the general equation is used:

$$v = \sqrt{(2gy + v_i^2)}$$

However, the displacement is negative from the maximum displacement until the starting point, at which time $y = 0$. From then on y has positive values.

Example

Continuing the example from above, where $v_i = -64$ ft/s, what are the velocities for various displacements on the way down?

Solution

The positive version of the formula is:

$$v = \sqrt{(64y + 4096)} \text{ ft/s}$$

Substitute values for y in the formula:

$y = -64$ ft	$v = 0$ ft/s	At peak or maximum displacement
$y = -32$ ft	$v = 45.3$ ft/s	Object moving downward
$y = 0$ ft	$v = 64$ ft/s	Moving downward at starting point
$y = 32$ ft	$v = 78.4$ ft/s	Moving downward below starting point

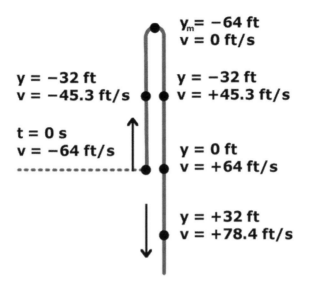

Velocities for displacements of object projected upward

Summary

When an object is projected upward, it is moving in the opposite direction of the force of gravity, and the initial velocity is a negative number.

The velocity is negative while the object moves up and positive while it moves downward. The displacement is negative above the starting point positive below the starting point. The equations for velocity are:

With respect to time

$$v = gt + v_i$$

With respect to displacement

$$v = -\sqrt{(2gy + v_i{}^2)} \text{ going up}$$

$$v_m = 0 \text{ (at maximum displacement)}$$

$$v = \sqrt{(2gy + v_i{}^2)} \text{ coming down}$$

Mini-quiz to check your understanding

1. If $v_i = -19.6$ m/s, what is the velocity at $t = 5$ s?

    a.   29.4 m/s

    b.   −29.4 m/s

    c.   9.8 m/s

2. If $v_i = -20$ ft/s, what is the velocity at $y = -2$ ft?

    a. $v = \pm\sqrt{(128 + 400)}$

    b. $v = -\sqrt{(128 - 400)}$ and $v = \sqrt{(-128 + 400)}$

    c. $v = \pm\sqrt{(-128 + 400)}$

3. If $v_i = -10$ m/s and $g = 9.8$ m/s^2, what is v when $y = 5$ m?

    a. $v = \pm\sqrt{(98 + 100)}$

    b. $v = \sqrt{(98 + 100)}$

    c. $v = \sqrt{(98 - 100)}$

Answers

1a, 2c, 3b

5.3 Displacement Equations for Objects Projected Upward

When you project an object upward and release it at some initial velocity, it travels upward until it reaches a maximum height, after which it falls toward the ground.

Since it is moving in the opposite direction of the force of gravity, its initial velocity is a negative number. Also, according to our convention for direction, the displacement above the starting point is negative, while the displacement below the starting point is positive.

Derived equations allow you to calculate the displacement with respect to velocity and with respect to elapsed time. You can also calculate the total distance the object travels from the starting point.

Displacement with respect to velocity

The general gravity equation for the displacement of an object with respect to velocity is:

$$y = (v^2 - v_i^2)/2g$$

where

- y is the vertical displacement in meters or feet
- v is the vertical velocity in m/s or ft/s
- v_i is the initial vertical velocity in m/s or ft/s; $v_i < 0$
- g is the acceleration due to gravity
  (9.8 m/s^2 or 32 ft/s^2)

(See chapter *2.4 Derivation of Displacement-Velocity Equations* for details of the derivation.)

When you project the object upward, it is moving in the opposite direction of gravity, and the initial velocity when you release it is negative or less than zero ($v_i < 0$).

On the way up

While the object is moving upward, the square of its velocity is less than the square of the initial velocity ($v^2 < v_i^2$). The result is the displacement from the starting point is negative ($y < 0$).

Maximum displacement with respect to velocity

At the peak or maximum displacement, the velocity is $v = 0$ and the displacement is:

$$y_m = (0 - v_i^2)/2g$$

$$y_m = -v_i^2/2g$$

where y_m is the maximum displacement.

On the way down

When the object is moving downward from the peak displacement but above the starting point, v^2 is still less than v_i^2 and the displacement is still negative ($y < 0$).

Once the object falls below the starting point, ($v^2 > v_i^2$) and the displacement becomes a positive value ($y > 0$).

Example

Suppose you throw a ball upward at 100 feet per second. What are the displacements from the starting point for the various velocities?

Solution

Since $v_i = -64$ ft/s, $g = 32$ ft/s^2. Substitute values for v_i and g into the equation:

$$y = (v^2 - v_i^2)/2g$$

$$y = [v^2 \text{ ft}^2/\text{s}^2 - (-64 \text{ ft/s})^2]/2*(32 \text{ ft/s}^2)$$

Combine and cancel out the units to get the formula:

$$y = (v^2 - 4096)/(64) \text{ ft}$$

Substitute in values for **v**:

$v = -64$ ft/s	$y = 0$ ft	At the starting point
$v = -32$ ft/s	$y = -48$ ft	Above starting point
$v = 0$ ft/s	$y_m = -64$ ft	Maximum displacement
$v = +32$ ft/s	$y = -48$ ft	Falling but above starting point
$v = +64$ ft/s	$y = 0$ ft	At the starting point
$v = +80$ ft/s	$y = +36$ ft	Below starting point

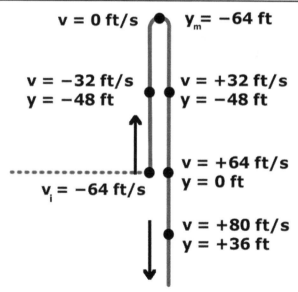

Displacements for velocities of object projected upward

Displacement with respect to time

The general gravity equation for the displacement of an object with respect to time is:

$$y = gt^2/2 + v_i t$$

where **t** is the time in seconds (s).

(See chapter *2.3 Derivation of Displacement-Time Equations* for details of the derivation.)

Since the initial velocity is negative, **y** will be negative for values of **t** when:

$$gt^2/2 < |v_i| t$$

where $|v_i|$ is the absolute or positive value of v_i. These are values of **t** where the object is above the starting point.

When $gt^2/2 > |v_i| t$, the displacement **y** is positive and the object is below the starting point.

Maximum displacement with respect to time

The equation for the maximum displacement with respect to time can be determined by starting with the equation:

$$t_m = -v_i/g$$

where t_m is the time to reach the maximum displacement

(See chapter *5.4 Time Equations for Objects Projected Upward* for more information.)

Solve for v_i:

$$v_i = -gt_m$$

Substitute for v_i in $y = gt^2/2 + v_i t$:

$$y_m = gt_m^2/2 - gt_m^2$$

$$y_m = -gt_m^2/2$$

Example

If the initial velocity is $v_i = -20$ m/s, what are the displacements for various times?

Solution

Substitute values for v_i and g in the equation:

$$y = gt^2/2 + v_i t$$

$$y = (9.8 \text{ m/s}^2)*(t^2 \text{ s}^2)/2 + (-20 \text{ m/s})*(t \text{ s})$$

Combine and cancel out units to get the formula:

$$y = (4.9t^2 - 20t) \text{ m}$$

Substitute in values for t. But also note that $t_m = -v_i/g$ and $y_m = -v_i^2/2g$:

$$t_m = 20/9.8 \text{ s} = 2.04 \text{ s}$$

$$y_m = -400/19.6 \text{ m} = 20.4 \text{ m}$$

$t = 0$ s	$y = 0$ m	At the starting point
$t = 1$ s	$y = -15.1$ m	Above starting point
$t_m = 2$ s	$y_m = -20.4$ m	At maximum height
$t = 3$ s	$y = -15.9$ m	Falling but above starting point
$t = 4$ s	$y = -1.6$ m	Nearing starting point
$t = 5$ s	$y = 22.5$ m	Below starting point

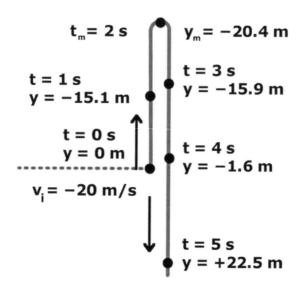

Displacements for various times of object projected upward

Total distance traveled

The displacement of an object is its movement in a specific direction from a starting point to some end point. It is a vector quantity.

Distance is a scalar quantity that is independent of direction and always has a positive value.

(See chapter *1.3 Convention for Direction in Gravity Equations* for more information.)

In the stated equations, the displacement **y** is negative above the starting point, when the object is going upward and downward. The value for **y** is positive below the starting point.

In order to find the total distance traveled, you need to state where the object is and then add the various displacements. In some cases, it is necessary to also know the maximum displacement.

Distance going up

The distance traveled of an object moving upward is simply the absolute value of the object's displacement:

$$d_u = |y|$$

where

- d_u is the distance going up
- $|y|$ is the absolute or positive value of the displacement going up

Total distance going up plus coming down

Total distance going up plus coming down

The total distance the object travels going up and then coming down, is the absolute value of two times the distance to the maximum displacement plus the displacement to the end point:

$$d = |2y_m| + y$$

where

- d is the total distance going up plus coming down
- $y_m = -v_i^2/2g$

If the object is above the starting point, y is negative and subtracted from $|2y_m|$. Below the starting point, everything is positive.

Example

Find the total distance traveled when the displacement is $y = -5$ m, $y_m = -10$ m and $y = +5$ m.

Going up

The total distance going up is:

$$d_u = |-5| \text{ m}$$

$$d_u = 5 \text{ m}$$

At maximum displacement

The total distance at the maximum displacement is:

$$\mathbf{d_m} = |-10| \text{ m}$$

$$\mathbf{d_m} = 10 \text{ m}$$

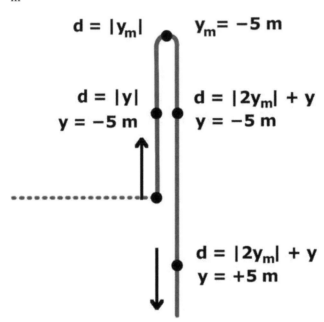

Total distances of object projected upward

Coming down, above starting point

The total distance coming down and above the starting point is:

$$\mathbf{d} = |-20| + (-5) \text{ m}$$

$$\mathbf{d} = 20 - 5 \text{ m}$$

$$\mathbf{d} = 15 \text{ m}$$

Coming down, below starting point

The total distance coming down and below the starting point is:

d = |−20| + (+5) m

d = 20 + 5 m

d = 25 m

Summary

When you project an object upward and release it at some initial velocity, it travels upward until it reaches a maximum displacement, after which it falls toward the ground.

The displacement above the starting point is negative, while the displacement below the starting point is positive.

Derived equations allow you to calculate the displacement for a given velocity, as well as the elapsed time for displacements both above and below the starting point. You can also calculate the total distance the object travels from the starting point. The equations for displacement are:

With respect to velocity

$y = (v^2 - v_i^2)/2g$

$y_m = -v_i^2/2g$ (maximum displacement)

With respect to time

$y = gt^2/2 + v_i t$

$y_m = -gt_m^2/2$ (maximum displacement)

Total distance traveled

$d_u = |y|$ (going upward to maximum displacement)

$d = |2y_m| + y$ (sum of going up and coming down)

Mini-quiz to check your understanding

1. If you double the initial velocity, how does that affect the maximum displacement?

    a. It is 4 times as great

    b. It is doubled

    c. It is cut in half

2. If the initial velocity is −10 ft/s, how far will the object travel in 1 second?

    a. 10 ft

    b. 6 ft

    c. 1 ft

3. If the maximum displacement is −20 m, what is $|2y_m|$?

    a. − 40 m

    b. It cannot be calculated

    c. 40 m

Answers

1a, 2b, 3c

5.4 Time Equations for Objects Projected Upward

When you project an object upward and release it at some initial velocity, it travels until it reaches a maximum height or displacement, after which it falls toward the ground.

The initial velocity has a negative value, the velocity is negative on the way up and positive on the way down, and the displacement is negative above the starting point and positive below the starting point.

Velocity and displacement are vectors, while time is a scalar quantity, which is always positive.

You can use derived equations to find the time it takes to reach a given velocity, as well as the time it takes to reach a given displacement, both above and below the starting point.

Time with respect to velocity

The general equation for the time it takes for an object to reach a given velocity is:

$$t = (v - v_i)/g$$

where

- t is the time in seconds (s)
- v is the vertical velocity in m/s or ft/s
- v_i is the vertical initial velocity in m/s or ft/s
- g is the acceleration due to gravity (9.8 m/s^2 or 32 ft/s^2)

(See chapter *2.2 Derivation of Velocity-Time Equations* for details of the derivation.)

When you project the object upward, it is moving in the opposite direction of gravity, and the initial velocity when you release it is negative or less than zero ($v_i < 0$).

On the way up

On the way up, v is negative ($v < 0$), and its absolute value is less than that of v_i:

$$|v| < |v_i|$$

where $|v|$ and $|v_i|$ are the absolute or positive values. Thus, $(v - v_i)$ has a positive value.

Time to reach maximum displacement

At the maximum displacement, $v = 0$ and the time equation becomes:

$$t_m = -v_i/g$$

where t_m is the time to reach the maximum displacement and $-v_i$ is a positive number.

On the way down

On the way down, $v > 0$ and $v < |v_i|$. Thus, $(v - v_i)$ has a positive value.

Example of times to reach various velocities

If $v_i = -128$ ft/s, find t for various velocities.

Solution

Since v_i is in ft/s, you use $g = 32$ ft/s^2.

Substitute values for v_i and g into the equation:

$$t = (v - v_i)/g$$

$$t = [(v \text{ ft/s}) - (-128 \text{ ft/s})]/(32 \text{ ft/s}^2)$$

Cancelling out units results in the formula:

$t = (v + 128)/32$ s

Substitute different values for **v** to get the various elapsed times:

v = −128 ft/s	t = 0 s	Starting off at initial velocity
v = −64 ft/s	t = 2 s	Moving upward
v = 0 ft/s	$t_m = 4$ s	At maximum displacement
v = 32 ft/s	t = 5 s	Moving downward
v = 128 ft/s	t = 8 s	Downward at starting point
v = 160 ft/s	t = 9 s	Below starting point

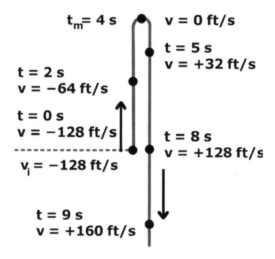

Times for various velocities of object projected upward

Time for displacements moving upward

The general gravity equation for elapsed time with respect to displacement is:

$$t = [-v_i \pm \sqrt{(v_i^2 + 2gy)}]/g$$

where

- ± means plus or minus
- $\sqrt{(v_i^2 + 2gy)}$ is the square root of the quantity $(v_i^2 + 2gy)$
- **y** is the vertical displacement in meters or feet

(See chapter *2.3 Derivation of Displacement-Time Equations* for details of the derivation.)

For an object projected upward, $-v_i$ is a positive number. In order for the object to start at **t** = 0 when **y** = 0, the equation for the time it takes an object to move upward toward the maximum displacement is the negative (−) version:

$$t = [-v_i - \sqrt{(v_i^2 + 2gy)}]/g$$

Time to reach maximum displacement

The equation for the time to reach maximum displacement with respect to the initial velocity has already been stated:

$$t_m = -v_i/g$$

The equation for the maximum displacement with respect to the initial velocity is:

$$y_m = -v_i^2/2g$$

(See chapter *5.3 Displacement Equations for Objects Projected Upward* for more information.)

These two equations make it easy to determine the time and maximum displacement. However, you may want to see the relationship between the factors.

Solve $y_m = -v_i^2/2g$ for v_i^2:

$$v_i^2 = -2gy_m$$

Square $t_m = -v_i/g$:

$$t_m^2 = v_i^2/g^2$$

Substitute in $v_i^2 = -2gy_m$:

$$t_m^2 = -2gy_m/g^2$$

$$t_m^2 = -2y_m/g$$

Take positive square root:

$$t_m = \sqrt{(-2y_m/g)}$$

Example for various displacements moving upward

If $v_i = -98$ m/s, find the times for various displacements as the object moves upward to the maximum displacement.

Solution

Since v_i is in m/s, $g = 9.8$ m/s^2. Also, since the displacements are above the starting point, the values of y will be negative numbers.

You can easily determine the time for the maximum height:

$$t_m = -v_i/g$$

$$t_m = 98/9.8 \text{ s}$$

$$t_m = 10 \text{ s}$$

Also:

$$y_m = -v_i^2/2g$$

$$y_m = -982/2*9.8 \text{ m}$$

$$y_m = -980/2 \text{ m}$$

$$y_m = -490 \text{ m}$$

You can verify those two values are correct by substituting in:

$$t_m = \sqrt{(-2y_m/g)}$$

$$10 = \sqrt{(-2*[-490]/9.8)}$$

$$10 = \sqrt{(100)} = 10$$

To find the various values of **t** with respect to **y**, substitute for $\mathbf{v_i}$ and **g** into the equation:

$$\mathbf{t = [-v_i - \sqrt{(v_i^2 + 2gy)}]/g}$$

You can independently verify that the units are correct and that **t** is in seconds:

$$\mathbf{t = [-(-98) - \sqrt{(-98^2 + 2*9.8y)}]/9.8} \text{ seconds}$$

One way to simplify the equation is by breaking up the fraction.

$$\mathbf{t = 98/9.8 - [\sqrt{(9604 + 2*9.8y)}]/9.8} \text{ s}$$

$$\mathbf{t = 10 - [\sqrt{(9604 + 19.6y)}]/9.8} \text{ s}$$

The equation still is not very simple, but it is workable. A simple case is when **y** = 0:

$$\mathbf{t = 10 - [\sqrt{(9604)}]/9.8} \text{ s}$$

$$\mathbf{t = 10 - 98/9.8} \text{ s}$$

$$\mathbf{t = 0} \text{ s}$$

The following results show the times for various displacements above and at the starting point:

y = 0 m	t = 0 s	At the starting point
y = −49 m	t = 0.5 s	Moving up and above starting point
y = −98 m	t = 1.1 s	Moving upward
y = −196 m	t = 2.25 s	Moving upward
y = −490 m	t = 10 s	At maximum displacement

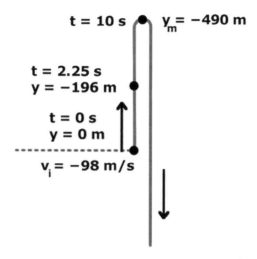

Times for various displacements in upward direction

Time for displacements moving downward

After the object reaches the maximum displacement and starts coming down, the equation changes to:

$$t = [-v_i + \sqrt{(v_i^2 + 2gy)}]/g$$

While the object is above the starting point, y is negative. When it reaches the starting point, $y = 0$. Then, when it is below the starting point, y becomes a positive number.

Example for various displacements moving downward

As in the previous example, if $v_i = -98$ m/s. Find the times for various displacements as the object moves downward.

Solution

You already know the maximum displacement and time:

$y_m = -490$ m

$t_m = 10$ s

Use the equation:

$$t = [-v_i + \sqrt{(v_i^2 + 2gy)}]/g$$

Substitute and simplify:

$$t = 10 + [\sqrt{(9604 + 19.6y)}]/9.8 \text{ s}$$

The following results show the times for various displacements as the object falls from the maximum height:

$y_m = -490$ m	$t_m = 10$ s	At maximum displacement
$y = -196$ m	$t = 17.75$ s	Falling
$y = -98$ m	$t = 18.9$ s	Falling but above starting point
$y = -49$ m	$t = 19.5$ s	Nearing starting point on way down
$y = 0$ m	$t = 20$ s	At starting point on the way down

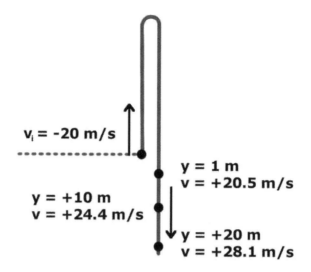

$v_i = -20$ m/s

$y = 1$ m
$v = +20.5$ m/s

$y = +10$ m
$v = +24.4$ m/s

$y = +20$ m
$v = +28.1$ m/s

Times for various displacements in downward direction

Summary

An object projected upward against the force of gravity slows down until it reaches a maximum displacement, after which its velocity increases as it falls toward the ground. Derived equations allow you to calculate the time it takes an object projected upward to reach a given velocity or a given displacement from the starting point.

Time with respect to velocity:

$$t = (v - v_i)/g$$

$$t_m = -v_i/g \text{ (At maximum displacement)}$$

The with respect to displacement:

$$t = [-v_i - \sqrt{(v_i^2 + 2gy)}]/g \text{ (going up)}$$

$$t_m = \sqrt{(-2y_m/g)} \text{ (At maximum displacement)}$$

$$t = [-v_i + \sqrt{(v_i^2 + 2gy)}]/g \text{ (coming down)}$$

Mini-quiz to check your understanding

1. If the initial velocity is −32 ft/s, how long does it take for the object to reach +32 ft/s?

    a. $t = (32 − 32)/32 = 0$ seconds

    b. $t = (32 + 32)/32 = 2$ seconds

    c. $t = 32 + 32/32 = 33$ seconds

2. If $v_i = −20$ m/s, which equation shows the time it takes the object to move upward and reach 30 m above the starting point?

    a. $t = [−20 − \sqrt{(−20)^2 + 2*9.8*30]/9.8}$

    b. $t = [20 + \sqrt{(−20)^2 + 2*9.8*(−30)]/9.8}$

    c. $t = [−20 − \sqrt{(−20)^2 + 2*9.8*(−30)]/9.8}$

3. If $v_i = −98$ m/s and $y = +196$ m, what is the elapsed time?

    a. 21.8 s

    b. −21.8 s

    c. 2.25 s

Answers

1b, 2c, 3a

Part 6: Gravity Applications

The principles of gravity and the various gravity equations for velocity, displacement and time can be applied to real world situations.

Part 6 Chapters

Chapters in Part 6 include:

6.1 Potential Energy of Gravity

This chapter gives the equations for the potential energy, kinetic energy and total energy of an object that is dropped, thrown downward or projected upward.

6.2 Work by Gravity Against Inertia

This chapter shows how the force of gravity can do work against the inertia of an object.

6.3 Work Against Gravity and Inertia by an External Force

An external force that lifts an object performs work against both gravity and the object's inertia.

6.4 Effect of Gravity on Sideways Motion

When you project an object sideways, it falls to the ground at the same rate, no matter what the initial sideways velocity, provided the velocity is not so great that the curvature of the Earth comes into play.

6.5 Effect of Gravity on an Artillery Projectile

This chapter explains the velocity components of a cannon projectile that is sent at an upward angle and shows how the distance is calculated, as well as the angle required for a given distance.

6.6 Gravity and Newton's Cannon

Isaac Newton's thought experiment concerning a "super-cannon" on the top of a very high mountain shows that, depending on the initial velocity, the cannonball will then hit the ground some distance away, go into an orbit around the Earth or fly off into space.

6.7 Escape Velocity from Gravity

This chapter shows how the escape velocity from gravity equation is determined and makes calculations of the escape velocities for the Earth, Moon and Sun. It also explains problems with using the equation.

6.8 Artificial Gravity

Artificial gravity is sometimes needed in space stations. This chapter shows how artificial gravity can be created and how fast a circular space station must rotate to simulate the Earth's gravity.

6.9 Center of Gravity

This chapter tells how you can find the center of gravity mathematically or experimentally.

6.1 Potential Energy of Gravity

An object held above the ground has the potential of accelerating downward, due to the pull of gravity. In other words, in that position, the object has potential energy (**PE**) that can be turned into the kinetic energy (**KE**) of motion.

The sum of the potential energy and kinetic energy due to gravity for an object is constant unless outside forces come into play.

You can calculate the **PE**, **KE** and total energy (**TE**) for an object that is dropped, thrown downward or projected upward with some simple equations. You can then verify that the final velocity is the same as obtained from the gravity derivations.

Energy for falling objects

An object held at a given height above the ground has an initial potential energy. When it is dropped, it gains kinetic energy. The total energy of the object is then used to find the velocity when the object hits the ground.

Potential energy for falling object

The equation for the object's initial **PE** with respect to gravity is::

$$PE_i = mgh$$

where

- PE_i is the initial potential energy in joules (J) or foot-pounds (ft-lbs)
- **m** is the mass of the object in kg-mass or pound-mass

- **g** is the acceleration due to gravity (9.8 m/s^2 or 32 ft/s^2)
- **h** is the height above the ground in m or ft

Note: Potential energy is also sometimes abbreviated as **U**.

When the object reaches the ground, **h** = 0 and thus the final potential energy is:

$$PE_f = 0$$

Note: In reality, there is still a gravitational force on the object at the surface of the Earth, so the object has a gravitational potential energy at that point. But since the object cannot go anywhere, we say its **PE** from gravity is zero.

Kinetic energy for falling object

Kinetic energy (**KE**) is the energy of motion. Since the object is not moving at the initial position, the initial **KE** is:

$$KE_i = 0$$

Once the object is released, it accelerates downward. When the object reaches the ground, its kinetic energy is:

$$KE_f = mv_f^2/2$$

where

- **KE$_f$** is the kinetic energy at the ground in joules (J) or foot-pounds (ft-lbs)
- **v$_f$** is the downward velocity of the object at the ground in m/s or ft/s

Total energy for falling object

The total energy of the object is:

$$T = PE + KE$$

The total energy is a constant value, provided no external forces besides gravity act on the object. Thus, the initial total energy equals the final total energy:

$$T_i = T_f$$

$$PE_i + KE_i = PE_f + KE_f$$

When the object is simply dropped,

$$mgh + 0 = 0 + mv_f^2/2$$

$$mgh = mv_f^2/2$$

Final velocity for falling object

From that equivalence, you can determine the final velocity of the dropped object. Divide by **m** and multiply by 2:

$$v_f^2 = 2gh$$

$$v_f = \sqrt{(2gh)}$$

This is equivalent to $v = \sqrt{(2gy)}$ that is given in the *Velocity Equations for Falling Objects* chapter.

Energy for objects projected downward

When an object at some height in projected downward and released, its initial velocity becomes a factor in the **KE**. However, it does not affect the **PE**.

$$PE_i = mgh \quad \bigcirc \quad KE_i = mv_i^2/2$$

$$\downarrow v_i$$

$$PE_f = 0 \quad \bigcirc \quad KE_f = mv_f^2/2$$

Initial and final PE and KE

Potential energy when projected downward

The **PE** of an object is independent of its velocity. It is only dependent of the height above the ground. Thus, the initial **PE** is:

$$PE_i = mgh$$

The final potential energy is:

$$PE_f = 0$$

Kinetic energy when projected downward

The initial kinetic energy of an object projected downward is dependent on its initial velocity:

$$KE_i = mv_i^2/2$$

where

- **KE$_i$** is the initial kinetic energy in joules (J) or foot-pounds (ft-lbs)
- v_i is the initial velocity of the object in m/s or ft/s

The object accelerates until it hits the ground at a final **KE**:

$$KE_f = mv_f^2/2$$

Total energy and final velocity

The total energy relationship is:

$$PE_i + KE_i = PE_f + KE_f$$

$$mgh + mv_i^2/2 = 0 + mv_f^2/2$$

Divide by **m**, multiply by 2 and rearrange terms to get the final velocity:

$$v_f^2 = 2gh + v_i^2$$

$$v_f = \sqrt{(2gh + v_i^2)}$$

This equation corresponds with the equation from *4.2 Velocity Equations for Objects Projected Downward*:

$$v = \sqrt{(2gy + v_i^2)}$$

where y is the displacement below the starting point.

Energy for objects projected upward

When an object is projected upward from a given height, it travels until it reaches a maximum displacement, at which time its velocity is zero. The object then falls to the ground from that displacement.

Potential energy when projected upward

The initial potential energy of an object projected upward is:

$$PE_i = mgh$$

As the object is moving upwards, the **PE** increases according to the displacement it reaches. The maximum displacement is::

$$y_m = -v_i^2/2g$$

where y_m is the maximum displacement from the starting point.

> (See chapter *5.3 Displacement Equations for Objects Projected Upward* for more information.)

> **Note**: The value of y_m is a negative number, because the motion is in the opposite direction of gravity.

However, the object is projected upward from a given height. Thus, the maximum height above the ground (h_m) that it reaches is:

$$h_m = h - y_m$$

The equation for the **PE** at the maximum height is then:

$$PE_m = mgh_m$$

or

$$PE_m = mgh - mgy_m$$

Substituting for y_m:

$$PE_m = mgh + mv_i^2/2$$

Considerations concerning PE_m can be made from the initial height.

Kinetic energy when projected upward

The initial **KE** is:

$$KE_i = mv_i^2/2$$

At the maximum displacement, $v = 0$ and thus:

$$KE_m = 0$$

When the object falls and finally reaches the ground:

$$KE_f = mv_f^2/2$$

Total energy and final velocity

To determine the final velocity, consider the total energy at the maximum displacement and compare it with the total energy at the ground:

$$PE_m + KE_m = PE_f + KE_f$$

$$(mgh + mv_i^2/2) + 0 = 0 + mv_f^2/2$$

Divide by **m**, multiply by 2 and rearrange terms to get the final velocity:

$$v_f^2 = 2gh + v_i^2$$

$$v_f = \sqrt{(2gh + v_i^2)}$$

This compares with the equation in *Velocity Equations for Objects Projected Upward* chapter:

$$v = \sqrt{(2gy + v_i^2)}$$

where **y** is the displacement below the starting point.

Summary

Potential energy with respect to gravity is **PE = mgh**.

When the object is dropped, thrown downward or projected upward, its kinetic energy becomes **KE = mv^2/2**, along with a factor of the initial velocity.

The sum of the **PE** and **KE** is the total energy, which is a constant. Equating the initial total energy with the final total energy, you can determine the final velocity of the object.

Mini-quiz to check your understanding

1. When an object is dropped from 5 m, what is its **PE** at 3m?

    a. **PE** = 5mg J

    b. **PE** = 3mg J

    c. **PE** = 0 J

2. When you project a 2 kg object downward, at 4 m/s, what is its initial **KE**?

    a. $KE_i = 16$ J

    b. $KE_i = 8$ J

    c. $KE_i = PE_i$

3. Does a dropped object have a greater final velocity than one that is projected upward?

    a. Yes, because the object projected upward goes in the opposite direction

    b. They have the same final velocity if the mass is the same

    c. No, because the object projected upward falls from a greater distance

Answers

1b, 2a, 3c

6.2 Work by Gravity Against Inertia

The amount of work is the force required to move an object some displacement against a resistance. It is the product of the force and the displacement caused by that force.

The inertia of matter is one such resistance to motion. When an object is dropped or projected downward, gravity does work to overcome the natural inertia of matter.

Thus, the product of the force of gravity and the displacement is the work done by gravity against inertia.

Work is also defined as the change in mechanical energy of the object as it moves from one position to another.

In the case of the force of gravity, work can be measured as the change in the potential energy or the change in the kinetic energy of the object. Equating the change in energies allows you to calculate the final velocity of the object after it moves a given displacement.

When the object is projected upward, the work above the starting point is negative and equals zero when the object returns to the starting point.

Work as force times displacement

A force is required to overcome the resistance of inertia and accelerate an object. As long as the force is being applied, the object will accelerate and work will be done against inertia.

The product of the displacement of the object and the force applied equals the work done against inertia.

Note: You may often see the word *distance* used in work. To be scientifically correct, *displacement* should be used instead. Distance can follow any path, while displacement is a vector and straight path in the line of the force.

(See chapter *1.3 Convention for Direction in Gravity Equations* for more information.)

The force of gravity to accelerate an object is:

F = mg

where

- **F** is the force of gravity in newtons (N) or pound-force (lbs)
- **m** is the mass of the object in kilograms (kg) or pound-mass (lbs)
- **g** is the acceleration due to gravity (9.8 m/s² or 32 ft/s²)

Note: Pounds are typically considered units of force or weight. However, some people also use the expression "pound" when referring to mass.

Thus, the unit of pound-force is used to distinguish it from pound-mass. Also, since **F = mg**, 1 pound-mass equals 32 pound-force.

Thus, the work done by gravity to overcome inertia is:

W = Fy

W = mgy

where

- **W** is the work done in joules (J) or pound-feet
- **y** is the vertical displacement from the starting point to some end point in m or ft

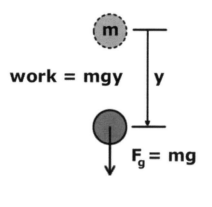

Work by gravity against inertia

Work as change in energy

The amount of work done by gravity to overcome the resistance of inertia can also be defined as either the change in potential energy (**PE**) or as the change in kinetic energy (**KE**) over the displacement:

$$W = \Delta PE$$

and

$$W = \Delta KE$$

where Δ is the Greek letter delta, indicating a change or difference.

Work as change in potential energy

The equation for potential energy of gravity is:

$$PE = mgh$$

where

- **PE** is the potential energy in joules (J) or foot-pounds (ft-lbs)
- **h** is the height above the ground in m or ft

(See chapter *6.1 Potential Energy of Gravity* for more information.)

The change in potential energy is:

$$\Delta PE = mgh_i - mgh_f$$

where

- h_i is the initial height from the ground
- h_f is the final height from the ground

Since **y** is the displacement the object falls from the starting point above the ground:

$$y = h_i - h_f$$

Multiplying both sides of equation by **mg**:

$$mgy = mgh_i - mgh_f$$

Thus:

$$mgy = \Delta PE$$

$$W = \Delta PE = mgy$$

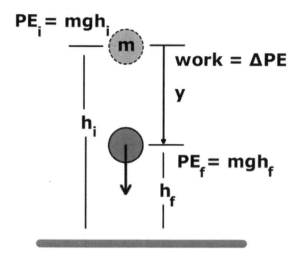

Work as change in potential energy

Work as change in kinetic energy

The equation for kinetic energy is:

$$KE = mv^2/2$$

The change in kinetic energy is:

$$\Delta KE = mv_f^2/2 - mv_i^2/2 = W$$

where

- v_f is the final velocity
- v_i is the initial velocity

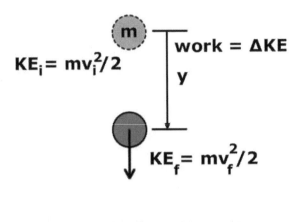

$$KE_i = mv_i^2/2 \qquad \text{work} = \Delta KE$$
$$y$$
$$KE_f = mv_f^2/2$$

Work as change in kinetic energy

Velocity for given displacement

Since $\Delta PE = \Delta KE$, you can find the final velocity for work done against inertia moving a given displacement:

$$mgy = mv_f^2/2 - mv_i^2/2$$

Divide by **m** and multiply by 2:

$$2gy = v_f^2 - v_i^2$$

Rearrange and take the square root:

$$v_f = \sqrt{(2gy + v_i^2)}$$

This is the same equation for the velocity of an object projected downward.

(See chapter *4.2 Velocity Equations for Objects Projected Downward* for more information)

When an object is simply dropped, $v_i = 0$ and the equation becomes:

$$v_f = \sqrt{(2gy)}$$

(See chapter *3.2 Velocity Equations for Falling Objects* for more information.)

Work when object projected upward

The work by gravity against inertia when an object is projected upward only occurs when the object starts falling downward. On the way up, you are doing work against gravity.

(See chapter *6.3 Work Against Gravity and Inertia by an External Force* for more information.)

The work done depends on whether you measure the work from the peak or maximum displacement or from the starting point where the object was released.

Work measured from maximum displacement

When an object is projected upward at some initial velocity, it will reach a maximum displacement before falling downward and doing work against inertia.

Note: The initial velocity is the velocity at which the object is released after being accelerated from zero velocity. Initial velocity does not occur instantaneously.

The equation for the maximum displacement is:

$$y_m = -v_i^2/2g$$

where y_m is the maximum displacement from the starting point in meters (m) or feet (ft)

Note: According to our convention for directions, displacements above the starting point are negative and thus $y_m < 0$. Also, upward velocities are negative and thus $v_i < 0$.

(See chapter *1.3 Convention for Direction in Gravity Equations* for more information.)

Work measured from the maximum displacement is simply work done by a falling object:

$$W = mgy$$

Work measured from starting point

When you project an object upward, you are doing work against gravity as a result of your initial velocity. Once the object starts falling downward, you can begin to measure the work gravity does to overcome inertia.

Above starting point

While the object is moving downward above the starting point, the work done by gravity with respect to the starting point is negative:

$$W = -mgy$$

Since the direction of **y** is in the opposite direction of gravity, it is a negative number, according to our convention for directions.

You can also see that Δ**PE** is negative, since $\mathbf{h}_f > \mathbf{h}_i$.

Likewise, Δ**KE** is negative, since $\mathbf{v}_f^2 < \mathbf{v}_i^2$.

Below starting point

As the object travels below the starting point, the work done by gravity is the same as if the object had been projected downward at a positive value of the initial velocity.

Summary

The product of the force of gravity and the displacement moved is the work done by gravity against inertia. Work is also the change in the potential energy or the change in the kinetic energy of the object.

Equating the change in energies allows you to calculate the final velocity of the object after it moves a given displacement.

When the object is projected upward, the work above the starting point is negative and equals zero when the object returns to the starting point. Afterwards, it follows the standard equations.

Mini-quiz to check your understanding

1. How much work is done by gravity when a 1 pound-mass ball is dropped 2 feet?

   a. 2 foot-pounds

   b. 64 foot-pounds

   c. 2 newtons

2. If the initial kinetic energy is 2 J and the final KE is 6 J, how much work is done?

   a. 4 J

   b. 12 J

   c. Not enough information to solve problem

3. How much work is done if an object is moving upward and gravity pulls it back to its starting point?

   a. **$mv^2/2$**

   b. No work is done

   c. **2mgy**

Answers

1b, 2a, 3b

6.3 Work Against Gravity and Inertia by an External Force

Work against gravity is achieved by applying a sufficient external force to move an object a certain displacement in the opposite direction of gravity. The work is the product of the force and the displacement.

If the object is initially stationary, the applied upward force must overcome both inertia and gravity to start to move the object.

Once the object is moving upward at some velocity, a force only equal to that from gravity is necessary to continue the upward movement at that velocity.

There are two common situations for determining the work required. You can project an object upward to a given height, where you let it continue to move—similar to throwing an object upward.

In the other situation, you move the object upward to a given velocity, keep it at that velocity and then cause it to slow down to zero velocity—as done when you lift something.

Work

Work is the effort over a displacement against a force that is resisting motion or is pulling in the opposite direction that you want to move the object.

In other words, work equals the product of the force that overcomes the resistance and the displacement in the same direction as the force:

$$W = Fd$$

where

- **W** is the work done
- **F** is the force applied against a resistive force
- **d** is the displacement in the same direction as the force

Although **F** and **d** are vector quantities with an indicated direction, **W** is a scalar quantity, with only magnitude and no direction.

Work against gravity

The force of gravity resists motion in its opposite direction. If an upward force equal to the force of gravity—or the weight of an object—is applied to a stationary object, the forces equal out, and the object does not move.

However, if the object has an initial upward velocity and a force equal to gravity is applied, the object will continue to move upward at that initial velocity.

The upward force is:

$$-F_g = m(-g_u)$$

$$F_g = mg_u$$

where

- $-F_g$ is the upward force needed to counter the force of gravity in newtons or pounds-force
- **m** is the mass of the object in kilograms or pounds-mass
- $-g_u$ is the acceleration in the opposite direction of the acceleration due to gravity (-9.8 m/s^2 or -32 ft/s^2)

Note: According to our convention for direction in gravity equations, **F**$_g$ and **g**$_u$ are negative numbers, since they are in the opposite direction of gravity.

The work done in moving an object against gravity to a certain height at an initial upward velocity is:

$$W = (-F_g)*(-y)$$

$$W = F_g y$$

where

- **W** is the work done against gravity in joules (J) or foot-pounds-force
- **−y** is the vertical displacement in meters (m) or feet (ft), measured from the starting point to when the force is discontinued

Note: Our convention states that **y** is negative when it is in the opposite direction of gravity.

Thus:

$$W = m(-g_u)(-y)$$

$$W = mg_u y$$

Note: Most physics textbooks use **h** for height, when talking about work from gravity: **W = mgh**. However, **h** is the displacement that an object is lifted above the ground, while **y** is the displacement from some starting point at or above the ground.

Work against inertia

Inertia is a resistance to changing the motion of an object. Its equation is:

$$-F_i = m(-a)$$

$$F_i = ma$$

where

- **−F_i** is the force required to overcome the inertia in newtons (N) or pounds-force (lbs)
- **m** is the mass of the object in kilograms (kg) or pounds-mass
- **−a_u** is the acceleration of the object in m/s² or ft/s²

The work against inertia in accelerating an object a displacement upward is:

$$W = F_i y$$

$$W = ma_u y$$

Note: At this point, we are not considering accelerating the object against the force of gravity. This equation is for the general work against inertia.

Work in projecting an object upward

When you project an object upward, you must not only overcome the force of gravity, but you must also overcome the resistance of inertia.

The force to project an object upward is a sum of the force needed to overcome gravity and the force required to overcome the object's inertia:

$$-F = -F_g - F_i$$

Work

If you apply those forces over the displacement, the amount of work is:

$$W = W_g + W_i$$

$$W = mg_u y + ma_u y$$

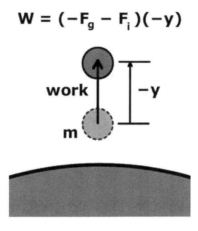

$$W = (-F_g - F_i)(-y)$$

Work against gravity and inertia

Object continues upward

However, once you stop applying the accelerating force to overcome inertia, the object will continue to move upward, due to its momentum. The velocity at the point where the force was stopped will be the initial velocity for an object projected upward.

> (See chapter *5.1 Overview of Gravity Equations for Objects Projected Upward* for more information.)

Example: Shot putter throws a lead ball

Suppose a shot putter throws the 16-pound lead ball straight up in the air. In accelerating the weight, his hand moves 2.5 feet in 1 second until the ball leaves his hand. How much work must he do to throw the ball?

Answer

The weight of the ball is 16 lbs, so its mass is 16/32 or 0.5 pounds-mass. The work against gravity is:

$$W_g = mg_u y$$

$$W_g = (0.5 \text{ pound-mass})*(-32 \text{ ft/s}^2)*(-2.5 \text{ ft})$$

$\mathbf{W_g}$ = 40 foot-pounds-force

Now, consider the acceleration of the ball to find the work to overcome inertia. Since the ball traveled 2.5 feet in 1 second, its average speed is 2.5 ft/s.

However, since the ball is starting at 0 ft/s, the end speed must be 5 ft/s to have an average of 2.5 ft/s. Thus, the acceleration is 5 ft/s².

The mass of the ball 0.5 pounds-mass, and the work done against inertia is:

$\mathbf{W_i = ma_u y}$

$\mathbf{W_i}$ = (0.5 pounds-mass)*(−5 ft/s²)*(−2.5 ft)

$\mathbf{W_i}$ = 6.25 foot-pounds-force

The total work is:

$\mathbf{W = W_g + W_i}$

$\mathbf{W}$ = (40 + 6.25) foot-pounds

$\mathbf{W}$ = 46.25 foot-pounds-force

You can see that the he must do extra work against inertia in this situation.

Work when ending at zero velocity

When you lift a stationary object, you typically accelerate it to some velocity and then lift it near the desired height, at which time you slow your lifting effort until the object has zero velocity.

A common example is lifting an object off the floor to place on a table.

Thus, the work done happens in three steps:

1. First, you do work against gravity and inertia until you reach a given velocity.

2. Then, you do work only against gravity alone, by lifting at a constant velocity

3. Finally, you do work against gravity but provide negative work against inertia in slowing the velocity to back to zero.

From zero to a given velocity

The work done in moving an object from zero to some given velocity is:

$$W_1 = mg_u y_1 + ma_u y_1$$

where

- W_1 is the work against gravity and inertia for the first step in the process
- y_1 is the vertical displacement moved until the object reaches a given velocity

Moving at constant velocity

The work done in moving an object at constant velocity is:

$$W_2 = mg_u y_2$$

where

- W_2 is the work against gravity for the second step in the process
- y_2 is the vertical displacement moved until near the desired height

Moving from some velocity to zero

The work done in moving an object from some given velocity to zero is:

$$W_3 = mg_u y_3 - ma_u y_3$$

where

- W_3 is the work against gravity with negative work against inertia for the third step in the process

- y_3 is the vertical displacement moved until the object reaches a zero velocity

What this means is that the force you are applying is less than the force required to lift the object, because you are overcoming its inertia in reducing its velocity to zero.

Total work

The total work is:

$$W = W_1 + W_2 + W_3$$

$$W = mg_u y_1 + ma_u y_1 + mg_u y_2 + mg_u y_3 - ma_u y_3$$

To simplify things, let's assume that the displacements for acceleration and deceleration are the same. Thus, the total work is:

$$W = mg_u y_1 + mg_u y_2 + mg_u y_3$$

$$W = mg_u y$$

where **y** is the total desired height or displacement.

> **Note** that although that this equation or **W = mgh** is given in most Physics textbooks, there seldom is mentioned the fact that the object must be accelerated and decelerated to reach the desired height.
>
> That fact is important for understanding of the principles involved.

The work done when the final velocity is zero is independent of the work against inertia, because negative work cancels out the positive work.

Example: Putting a heavy box on a shelf

You want to lift a 34 kg-force (75 pound-force) weight up to a shelf that is 2 meters off the ground.

You accelerate the box from being stationary to a velocity of 2 m/s, lift it up and then decelerate the box back to 0 m/s when you place it on the shelf. How much work must you do?

Answer

Since you are lifting the box from zero velocity to zero velocity, the work against inertia cancels out, and you are only doing work against gravity.

The weight of the ball is 34 kg-force, so its mass is 34/9.8 or 3.47 kg-mass. The work against gravity is:

$$W_g = mg_u y$$

$$W_g = (3.47 \text{ kg-mass})*(-9.8 \text{ m/s}^2)*(-2 \text{ m})$$

$$W_g = 68 \text{ joules}$$

Summary

Work against gravity and inertia is achieved by applying a sufficient external force to accelerate an object a certain displacement in the opposite direction of gravity.

The work is the product of the applied force and the displacement.

When an object is thrown upward, you must overcome both gravity and inertia. Assuming a constant accelerating force, the work done is $W = mg_u y + ma_u y$.

When the object is simply lifted to a height, at which point its velocity is zero, the work done is $W = mg_u y$.

Mini-quiz to check your understanding

4. How much force is needed to project an object upward?

    a. More than its weight

    b. Equal to its weight

    c. Less than its weight

5. Why is inertia not a factor in lifting an object onto the table?

    a. It is only a factor when dropping an object

    b. Work in slowing the object to zero cancels out the work against inertia in speeding it up

    c. It depends on how high the table is

6. How much work is required to lift a 2 kg weight from the ground to a height of 1 meter?

    a. 2 foot-pounds

    b. 2 newtons

    c. 2 joules

Answers

1a, 2b, 3c

6.4 Effect of Gravity on Sideways Motion

When an object is moving sideways, horizontal or parallel to the Earth's surface at a constant velocity, the effect of gravity on the object is independent of the object's lateral movement.

In other words, an object moving sideways will fall at the same rate as one that is simply dropped. You can derive the equation for the displacement of the object before it hit the ground from the gravity equations for falling objects.

An exception is if the object moves so far that the curvature of the Earth comes into play during its fall to the ground.

Gravity does not affect sideways motion

The force of gravity acts on objects in a direction that is perpendicular to level ground.

> **Note**: Since the Earth is a sphere with a circumference of approximately 40,000 km or 25,000 mi, the ground can be considered level or flat (not counting hills and valleys) for short displacements of several kilometers or several miles.

This means that, if an object is moving parallel to the ground, the force of gravity is only pulling on the object in a downward direction. The force is not affected by sideways motion and simply pulls the object down at the same rate as if it was stationary.

> (This rule is explained in detail in chapter *1.7 Horizontal Motion Unaffected by Gravity*.)

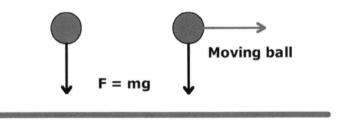

Gravity pull is same for moving and stationary objects

Objects hit ground at same time

If you would project or throw an object exactly parallel to the Earth's surface, the sideways motion of the object would have no effect on how gravity acts on it. In other words, the object would drop at the same rate as an object dropped from the identical height. The time it would take either object to hit the ground would be the same.

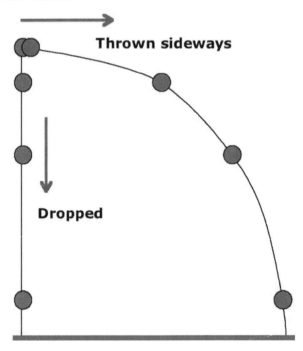

Ball thrown sideways falls at same rate as dropped ball

Simple experiment

You can try a simple experiment to verify this phenomenon. Place a coin on the edge of a table or desk and hold another coin at the same height. With one hand flick the coin on the table across the room. At the same time, drop the other coin. You will hear that they hit the floor at just about the same time.

Derivation of equation for displacement

You can find the displacement an object projected sideways from the following derivation. First, consider the displacement-velocity equation:

$$x = v_s t$$

where

- x is the horizontal displacement in feet (ft) or meters (m)
- v_s is the initial sideways velocity in ft/s or m/s
- t is the time until the object hits the ground in seconds (s)

The equation for the time a falling object takes is:

$$t = \sqrt{(2y/g)}$$

where

- y is the height in ft or m
- g is the acceleration due to gravity (32 ft/s^2 or 9.8 m/s^2)

(See chapter *3.4 Time Equations for Falling Objects* for information on the equation.)

Thus, the displacement the object travels, as a function of the initial velocity and the height is:

$$x = v_s \sqrt{(2y/g)}$$

Shooting and dropping a bullet

If you would shoot a bullet from a gun exactly parallel to the Earth's surface, the motion of the bullet would have no effect on how gravity acts on the bullet. In other words, the bullet would drop at the same rate as a stationary object.

Dropped bullet and shot bullet hit ground at same time

Many people don't believe that if you held a rifle or handgun parallel to the ground and at the same time you shot the bullet, you dropped another bullet from the same height, both bullets would both hit the ground at the same time.

However, it is a fact.

Exception

An exception to this phenomenon would be if the bullet or object was able to travel so many miles or kilometers that the curvature of the Earth came into play.

In such a situation, the bullet would take slightly longer to hit the ground, because the displacement to the ground was greater due to the Earth's curvature.

Example

If you shot a bullet at 900 m/s from a rifle that was 1.5 m above the ground, how far would the bullet fly until it hit the ground? Discount air resistance and assume the rifle is parallel to the ground.

Solution

$$x = v_s \sqrt{(2y/g)}$$

$$x = (900 \text{ m/s}) \sqrt{[2*(1.5 \text{ m})/(9.8 \text{ m/s}^2)]}$$

$$x = (900 \text{ m/s}) \sqrt{(0.306 \text{ s}^2)}$$

$$x = (900 \text{ m/s})(0.553 \text{ s})$$

$$x = 498 \text{ m or } 1634 \text{ ft}$$

Summary

An object moving sideways or parallel to the Earth's surface will fall at the same rate as one that is simply dropped.

The equation for the displacement of the object before it hits the ground can be derived from the gravity equations for falling objects.

An exception is if the object moves so fast or far that the curvature of the Earth comes into play during its fall to the ground.

Mini-quiz to check your understanding

1. Why doesn't gravity affect an object moving sideways differently?

   a. Gravity pulls perpendicular to the ground, so sideways motion is not affected

   b. Sideways motion is too fast for gravity to affect it

   c. Gravity pulls sideways motion faster than stationary motion

2. After 2 seconds, which object is falling downward faster?

   a. The stationary object falls faster because gravity affects it more

   b. The object moving sideways falls downward faster because of less friction

   c. They both are moving at the same downward velocity after 2 seconds

3. If two balls are thrown sideways at different velocities, will they hit the ground at the same time?

   a. Yes, because sideways velocities do not affect the rate the balls fall

   b. No, because the harder you throw a ball, the less gravity affects it

   c. It depends whether the balls are the same weight or not

Answers

1a, 2c, 3a

6.5 Effect of Gravity on an Artillery Projectile

The artillery in the military uses cannons or howitzers to send explosive projectiles into enemy territory. Artillery personnel determine the horizontal displacement of the target and adjust the angle of the cannon according to the known initial velocity of the projectile. It is assumed that the effect of air resistance is negligible on the projectile.

> **Note**: Horizontal displacement is how far the projectile has traveled from the cannon along the horizontal axis. It is a vector in a specified direction. Distance is a scalar quantity that indicates how far the object traveled along its curved path.
>
> (See chapter *1.3 Convention for Direction in Gravity Equations* for more information.)

Artillery equations start with the initial velocity of the projectile, which can be divided into its vertical and horizontal components. This results in the vertical velocity component following the equations for an object projected upward and the horizontal velocity component following the simple displacement equation.

The method to calculate the horizontal displacement the projectile is to determine the time it takes for the projectile to reach its maximum height and return to the ground and then multiply that time by the horizontal component of the velocity.

The equation for the angle required to achieve the desired displacement comes from solving the displacement equation as a function of the angle.

Velocity components

A projectile shot from a cannon starts with an initial or muzzle velocity of $-v_i$ at an angle θ (Greek letter theta) with respect to the ground. It leaves the cannon at a displacement **h** from the muzzle to the ground. These variables are used to determine the horizontal displacement of the projectile.

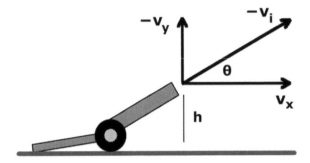

Projectile leaves cannon at angle θ with the ground

The initial velocity of the projectile can be separated into its **x** and **y** components, where v_x is the initial velocity in the **x** or horizontal direction and $-v_y$ is the initial velocity in the **y** or vertical direction.

Note: According to our convention for direction, the vector v_x has a positive value and both v_i and v_y are negative.

(See chapter *1.3 Convention for Direction in Gravity Equations* for more information.)

Components of initial velocity

The **x** and **y** components of the projectile initial velocity are functions of the angle of the cannon:

$$v_x = -v_i\cos(\theta)$$

where

- v_i is the initial velocity vector
- v_x is the component of the initial velocity in the horizontal or **x** direction
- θ is the angle of the cannon with respect to the ground
- $\cos(\theta)$ is the cosine of the angle θ

and

$$v_y = v_i sin(\theta)$$

where

- v_y is the component of the initial velocity in the vertical or **y** direction
- $sin(\theta)$ is the sine of the angle θ

Components act independently

According to the rule stated in the *Horizontal Motion Unaffected by Gravity* chapter, the perpendicular velocity components act independently of each other.

This means that the vertical motion of the projectile follows the gravity equations, with an initial velocity of v_y. The horizontal motion is simply a function of v_x and elapsed time.

Path of projectile

The projectile will follow a parabolic path as it moves upward until it reaches its maximum vertical displacement (y_m). It then falls to the ground. At the same time, the projectile is moving in the horizontal direction at velocity v_x.

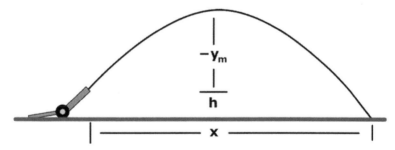

Path of projectile fired from cannon

The projectile leaves the cannon barrel from the height **h**, which presents an small added factor in calculating the time it takes to hit the ground, as well as the displacement the projectile travels.

Maximum vertical displacement

The maximum vertical displacement (y_m) of the projectile is:

$$y_m = -v_y^2/2g$$

(See chapter *5.3 Displacement Equations for Objects Projected Upward* for more information.)

Substitute $v_y = v_i sin(\theta)$ in the equation:

$$y_m = -v_i^2 * sin^2(\theta)/2g$$

Weight of projectile not a factor

The weight of the projectile determines the amount of propulsion material needed for the desired initial or muzzle velocity. However, neither weight nor mass is a factor in gravity equations for projected objects.

Determining displacement from cannon

The vertical component of the initial velocity acts as an initial velocity of an object projected upward. You can use the *Time Equations for Objects Projected Upward* chapter to determine the time it takes for the projectile to hit the ground.

You can then use the simple horizontal displacement equation $x = v_x t$ to determine how far the projectile will travel to its target.

Time to hit the ground

The projectile moves upward until it reaches a maximum vertical displacement. Then it falls to the ground. Since the barrel of the cannon is above the ground, that height should be added in.

The time it takes an object projected upward to reach the maximum vertical displacement and then fall to a displacement below the starting point is:

$$t = [-v_y - \sqrt{(v_y^2 + 2gh)}]/g$$

where

- **t** is the total time in seconds
- **g** is the acceleration due to gravity (9.8 m/s^2 or 32 ft/s^2)
- **h** is the displacement below the starting point in m or ft

(See chapter *5.4 Time Equations for Objects Projected Upward* for more information.)

Horizontal displacement of the projectile

The horizontal displacement of the projectile is then:

$$x = v_x t$$

$$x = v_x*[-v_y - \sqrt{(v_y^2 + 2gh)}]/g$$

$$x = -v_x v_y/g - v_x \sqrt{(v_y^2 + 2gh)}/g$$

Typically, the height of the end of the cannon, **h**, is very small compared to the horizontal displacement. You can simplify the displacement equation by setting **h** = 0, such that the equation reduces to:

$$x = -v_x v_y/g - v_x \sqrt{(v_y^2)}/g$$

Simplifying and combining terms:

$$x = -2v_x v_y/g$$

Put in terms of initial velocity and angle

You can put the simple horizontal displacement equation in terms of initial velocity and angle by substituting $v_x = -v_i\cos(\theta)$ and $v_y = v_i\sin(\theta)$ into the equation:

$$x = -2v_x v_y/g$$

$$x = -2[-v_i\cos(\theta)]*[v_i\sin(\theta)]/g$$

$$x = 2v_i^2\sin(\theta)*\cos(\theta)/g$$

Since $\sin(\theta)*\cos(\theta) = \sin(2\theta)/2$, the resulting horizontal displacement equation is:

$$x = v_i^2\sin(2\theta)/g$$

Example 1

If the cannon is at an angle of $\theta = 30°$, the end of the barrel is 4 ft off the ground and the initial velocity of the projectile is $v_i = 640$ ft/s, how far will the projectile travel?

Solution

Since the 4 feet adds little to the displacement, the simple equation can be used:

$$x = v_i^2\sin(2\theta)/g$$

Substitute values, with $g = 32$ ft/s^2:

$$x = 640^2*\sin(60°)/32 \text{ ft}$$

$$x = 409600*0.866/32 \text{ ft}$$

$$x = 11084.8 \text{ ft} = 2.1 \text{ miles}$$

Note: The 4 ft height would only add 5.5 feet to the horizontal displacement, which is insignificant at over 2 miles.

Example 2

Suppose the angle was set at $(90° - 30°) = 60°$. What would the displacement be?

Solution 2

Since $2*60° = 120°$ and $\sin(120°) = 0.866$, the displacement is:

$$x = 11084.8 \text{ ft} = 2.1 \text{ miles}$$

In other words, complementary angles 30° and 60° (90° − 30°) result in the same displacement.

This is true for any such pair of complementary angles:

θ and $(90° - \theta)$

It is something for you to work out.

Example 3

Examples 1 and 2 showed that the displacement for 30° and 60° is the same.

What is the difference in maximum height between the two angles?

Solution 3

For $\theta = 30°$:

$$y_m = -v_i{}^2\sin^2(\theta)/2g$$

$$y_m = -640^2{}^*\sin^2(30°)/2^*32$$

$$y_m = -409600^*0.25/64$$

$$y_m = -1600 \text{ ft}$$

For $\theta = 60°$:

$$y_m = -640^2{}^*\sin^2(60°)/64$$

$$y_m = -409600^*0.75/64$$

$$y_m = -4800 \text{ ft}$$

The 60° angle results in a greater maximum vertical displacement, but the horizontal displacement is the same as for 30°.

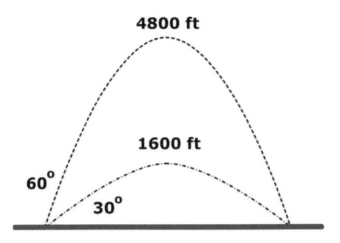

Same displacement for complementary angles

Determining angle to hit target

Soldiers using a mortar adjust the angle of the device to hit a target at a defined displacement.

Start with the displacement equation to derive the equation for the angle θ:

$$x = v_i^2 \sin(2\theta)/g$$

Solve for θ:

$$\sin(2\theta) = gx/v_i^2$$

The result has two possible solutions:

$$\theta = [\arcsin(gx/v_i^2)]/2$$

and

$$\theta = 90° - [\arcsin(gx/v_i^2)]/2$$

which are complementary angles.

Also, **arcsin(gx/v_i^2)** is the arcsine of **gx/v_i^2**. That means that **θ** is an angle that has a sine of **gx/v_i^2**. A scientific calculator is necessary to determine the arcsine of an angle.

Example

A target is seen at 500 m away. The muzzle velocity of the mortar is 160 m/s. What is the setting of the mortar angle?

Solution

Since the displacement is in meters, **g** = 9.8 m/s^2. **x** = 1000 m and **v_i** = −160 m/s.

$$\theta = [\text{arcsin}(gx/v_i^2)]/2$$

$$\theta = [\text{arcsin}(9.8*1000/(-160)^2)]/2$$

$$\theta = [\text{arcsin}(0.383)]/2$$

$$\theta = 22.5°/2 = 11.25°$$

However, that angle is too shallow to be practical. Instead, use 90° − **θ**:

$$90° - \theta = 78.75°$$

This is more of a typical angle for a mortar.

Summary

A cannon, howitzer or mortar sends projectiles a displacement away, as determined by the initial or muzzle velocity of the projectile and the angle of inclination.

The velocity of the projectile can be divided into its vertical and horizontal components. These velocity components are independent of each other.

Given the initial velocity and inclination angle, the equation for the horizontal displacement traveled is:

$$x = v_i^2 \sin(2\theta)/g$$

Two equations determine the necessary angle to hit a target at a given horizontal displacement:

$$\theta = [\arcsin(gx/v_i^2)]/2$$

$$\theta = 90° - [\arcsin(gx/v_i^2)]/2$$

Mini-quiz to check your understanding

1. Why is the muzzle velocity broken into perpendicular components?

    a. To simplify the mathematics when the height is small

    b. So that the velocities add up to be 90°

    c. Because the components act independently of each other

2. What angle would give you the maximum vertical displacement for a given initial velocity?

    a. 45°

    b. 90°

    c. 180°

3. Why would you use the larger complementary angle to project a certain horizontal displacement?

    a. To avoid hitting something that might be in the way

    b. Because the projectile will go further with the larger angle

    c. You should always use the smallest possible angle

Answers

1c, 2a, 3a

6.6 Gravity and Newton's Cannon

Newton's Cannon (also called *Newton's Cannonball*) was a "thought experiment" created by Isaac Newton in 1687 and stated in his *Principia Mathematica*.

In the book, he imagined shooting a cannonball parallel to the Earth's surface from the top of a very high mountain.

Depending on the velocity of the cannonball, it would strike the ground at some distance from the mountain top, go into orbit around the Earth or fly off into space. Of course, this assumes air resistance is negligible.

Cannonball hits the ground

When a cannonball is fired parallel to the Earth's surface from a cannon on the top of a high mountain, the ball will usually travel for some distance until it hits the ground.

Assuming air resistance is negligible, the cannonball's displacement depends on its initial horizontal velocity and the time it takes the ball to fall to the ground from its given height.

> **Note**: Displacement is how far the cannonball has traveled from the cannon along the horizontal axis. It is a vector in a specified direction. Distance is a scalar quantity that indicates how far the object traveled along its curved path.
>
> (See chapter *1.3 Convention for Direction in Gravity Equations* for more information.)

When curvature not a factor

If the velocity of the cannonball is only sufficient to carry it several kilometers or miles, the curvature of Earth does not really come into play.

The equation for its displacement is:

$$x = v_h \sqrt{(2y/g)}$$

where

- x is the horizontal displacement in meters (m) or feet (ft)
- v_h is the initial horizontal velocity in m/s or ft/s
- y is the displacement the cannonball has fallen from its starting point in m or ft
- g is the acceleration due to gravity (9.8 m/s^2 or 32 ft/s^2)

(See chapter *6.4 Effect of Gravity on Sideways Motion* for more information.)

Would travel in parabolic path

You can see that if the ground is considered flat that the cannonball travels in a parabolic path by squaring both sides of the equation:

$$x^2 = (2v_h{}^2/g)y$$

This results in the equation of a parabola, where the constant $k = 2v_h{}^2/g$:

$$x^2 = ky$$

When curvature is a factor

For greater distances, the curvature of the Earth becomes a factor and complicates the equation.

When the ground is considered flat, the force of gravity is perpendicular to the horizontal velocity of the cannonball. However, when the curvature of the Earth is taken into consideration, the direction of gravity changes with the distance traveled.

It is assumed that the force of gravity is concentrated at the center of the Earth.

The cannonball travels in an elliptical path that is interrupted by the surface of the Earth. The lower axis of the ellipse is at the center of the Earth.

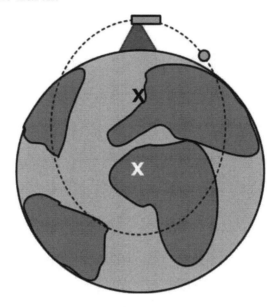

Cannonball follows elliptical path to hit the ground

As the initial velocity of the cannonball is increased, the ellipse becomes larger and the upper axis approaches the lower axis.

Cannonball goes into orbit

At some initial velocity, the cannonball will not hit the ground but will go into orbit around the Earth.

Initial elliptical orbit

As the initial velocity of the cannonball is increased, the size of the elliptical path increases to the point where the ball does not hit the ground and instead barely orbits the Earth. The upper axis of the ellipse approaches the lower axis at the center of the Earth.

Reaches circular orbit

At a certain initial velocity of the cannonball, the upper axis and lower axis of the ellipse coincide at the center of the Earth. In other words, the elliptical orbit becomes a circular orbit. The initial velocity for a circular orbit is:

$$v_c = \sqrt{(gR_E)}$$

where

- v_c is the initial velocity of the cannonball required for a circular orbit in m/s
- **g** is the acceleration due to gravity = 9.8 m/s^2
- R_E is the radius of the Earth = 6.371*10^6 m

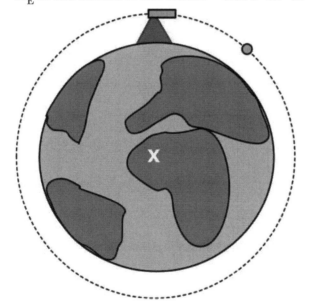

Cannonball goes into a circular orbit

Circular orbit example

What would be the required initial velocity of the cannonball to orbit the Earth?

Solution

Since the radius of the Earth is R_E = 6.371*10^6 m, the initial velocity is:

$$v_c = \sqrt{(9.8*6.371*10^6)}$$

$$v_c = \sqrt{(62.436*10^6)}$$

$$v_c = 7.902*10^3 \text{ m/s} = 7902 \text{ m/s}$$

This is much greater than the muzzle velocity of modern military cannons, which reach velocities of about 1,800 m/s. In fact the new electromagnetic railguns being developed by the U.S. Navy reach only 2400 m/s.

The result is that Newton's Cannon could work in theory, but there is no existing way to fire a cannonball or any projectile at the velocity required to orbit the Earth.

Elliptical orbit

As the initial velocity of the cannonball increases, the orbit becomes elliptical again. The upper axis of the ellipse is now at the Earth's center. The lower axis moves further away as the initial velocity is increased.

The velocity of the ball is for these elliptical paths greater than that for a circular path but less than the escape velocity:

$$\sqrt{(gR_E)} < v_c < \sqrt{(2gR_E)}$$

where

- $\sqrt{(gR_E)} < v_c$ means that v_c is greater than the velocity for a circular orbit
- $v_c < \sqrt{(2gR_E)}$ means that v_c is less than the escape velocity

(See chapter *12.6 Effect of Velocity on Orbital Motion* for more information.)

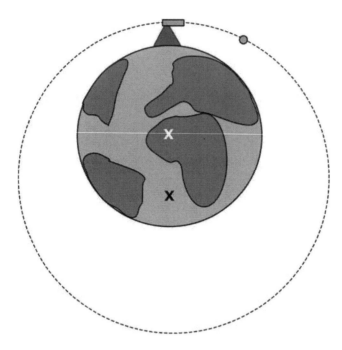

Cannonball goes into large elliptical orbit around Earth

Cannonball goes into space

The cannonball will fly off into space, if its initial velocity is:

$$v_e \geq \sqrt{(2gR)}$$

where

- v_e is the escape velocity from Earth's gravity
- $\geq$ means greater than or equal to...

If $v_e = \sqrt{(2gR_E)}$, the path of the cannonball is parabolic.

If $v_e > \sqrt{(2gR_E)}$, the path of the cannonball is hyperbolic.

(See chapter *12.6 Effect of Velocity on Orbital Motion* for more information.)

Summary

Newton's Cannon was a "thought experiment" created by Isaac Newton where he imagined shooting a cannonball parallel to the Earth's surface from the top of a very high mountain.

Depending on the velocity of the cannonball, it would strike the ground at some displacement from the mountain top, go into orbit around the Earth or fly off into space. The paths of the cannonball would be elliptical until the velocity reaches the escape velocity, where the path would become parabolic or hyperbolic.

However, the required velocities for orbiting the Earth or going into space are really larger than possible for even modern cannons.

Mini-quiz to check your understanding

1. Would Newton's Cannon have similar results on the Moon?

    a. No, because the Moon has no gravity

    b. It depends if a rocket could bring a cannon up to the Moon

    c. Yes, except you use different values of **g** and **R**

2. How would Newton increase the radius of the circular orbit?

    a. Shoot the cannonball from a higher mountain

    b. Shoot the cannonball straight up

    c. There is only one possible radius

3. How could a cannonball possibly fly off into space?

    a. Newton had some magical powers

    b. This is a thought experiment that follows laws of Physics

    c. A very small cannonball would be necessary

Answers

1c, 2a, 3b

6.7 Escape Velocity from Gravity

Escape velocity from gravity is the initial velocity of an object projected upward that is sufficient to overcome the downward pull of gravity and not fall back to the ground. Since the object is at or near the surface of the Earth, it is often called *surface escape velocity*.

The equation for the escape velocity comes from the gravitational escape velocity equation, as applied to objects near the Earth's surface. The resulting simple equation gives the escape velocity as a function of the acceleration due to gravity and the Earth's radius.

You can also apply the equation to the Moon and Sun, provided you know the radius and gravity of each of those bodies.

> Although the values for the surface escape velocity are commonly used in textbooks and scientific papers, they are unfortunately misleading and not realistic. There are several problems that prevent the escape velocity from gravity from being practical.

Determining escape velocity

The escape velocity from gravity can be obtained from the general gravitational escape velocity equation by substituting in the acceleration due to gravity.

This requires that the object is projected upward at some initial velocity from relatively close to the Earth's surface—typically, less than 65 km or 40 miles in altitude.

The equation does not apply to continuously propelled objects, such as rockets. Also, air resistance and the rotation of the Earth are not considered.

Gravitational escape velocity equation

The gravitational escape velocity equation is:

$$v_e = -\sqrt{(2GM/R_i)}$$

where

- v_e is the escape velocity in kilometers/second (km/s)
- G is the Universal Gravitational Constant = $6.674*10^{-20}$ km³/kg-s²
- M is the mass of the planet or sun in kilograms (kg)
- R_i is the initial separation between the center of mass of the planet or sun and the center of the object in kilometers (km)

Note: The negative sign in the equation indicates that the velocity vector is in a direction opposite of the gravitational force vector, according to our direction convention. Be aware that many textbooks state the equation as a positive value.

(See chapter *1.3 Direction Convention in Gravitation Equations* for more information.)

Also note: Since escape velocity is typically stated in km/s, the value of the Universal Gravitation Equation, **G**, is defined in km³/kg-s² and the separation **R** in km.

(See chapter *13.2 Gravitational Escape Velocity* for more information.)

Relationship with acceleration due to gravity

For objects on the surface of the Earth, the acceleration due to gravity is:

$$g = GM_E/R_E{}^2$$

where

- g is the acceleration due to gravity = 9.8 m/s² or 0.0098 km/s²
- M_E is the mass of the Earth = $5.974*10^{24}$ kg
- R_E is the radius of the Earth = 6371 km

Note: Since escape velocity is in km/s, **g** is stated in km/s² instead of m/s².

Also note: The value of **g** is considered constant at heights near the Earth's surface. At an altitude of about 64 km or 40 mi, the value of **g** reduces 2%, going from 9.8 m/s^2 to 9.6 m/s^2 and the assumption of being constant starts to fail.

(See chapter *1.4 Gravity Constant* for more information.)

Equation for escape velocity from gravity

To find the equation for the escape velocity from gravity, multiply both sides of the $\mathbf{g} = \mathbf{GM_E/R_E^2}$ relationship by $\mathbf{R_E}$:

$$\mathbf{gR_E = GM_E/R_E}$$

Substitute into $\mathbf{v_e = -\sqrt{(2GM/R)}}$ to get the equation for the escape velocity from gravity:

$$\mathbf{v_e = -\sqrt{(2gR_E)}}$$

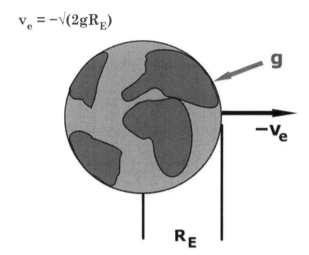

Factors for escape velocity from Earth's gravity

Escape speed

Although $\mathbf{v_e}$ is a vector that indicates the direction of the velocity as away from the Earth, it is more convenient to simply indicate the magnitude or speed of the escape velocity. Thus, the equation you usually see is:

$$\mathbf{s_e = \sqrt{(2gR_E)}}$$

where s_e is the escape speed or magnitude of the escape velocity vector v_e.

The escape from gravity equation is simpler to use than the gravitation equation, especially since the mass of the Earth is not necessary in calculations.

Escape velocity at an angle

If you project an object straight up, it will travel until it reaches its maximum height and then fall back to the ground. If you project it at an angle, it will follow a parabolic path until it hits the ground.

If the horizontal velocity is sufficient, the object will go into orbit around the Earth. If the vertical velocity is sufficient, the object will follow a curved path but never return to Earth.

> (Chapter *6.6 Gravity and Newton's Cannon* goes through the various possibilities.)

In other words, the angle does not affect the value of the escape velocity from gravity.

Common escape velocities

You can use the equation to determine escape velocity from the gravity of the Earth, Moon and Sun.

Earth

Since the radius of the Earth, R_E, is about 6371 km and $g = 0.0098$ km/s^2. The escape velocity is:

$$s_e = \sqrt{(2gR_E)}$$

$$s_e = \sqrt{[2*(0.0098 \text{ km/s}^2)*(6371 \text{ km})]}$$

$$s_e = \sqrt{(124.872 \text{ km}^2/\text{s}^2)}$$

$$s_e = 11.175 \text{ km/s}$$

Thus, the escape velocity from the surface of the Earth is about 11.2 km/s or 26,000 miles per hour. This is the same value calculated from $v_e = -\sqrt{(2GM/R)}$ at the Earth's surface.

Moon

The radius of the Moon is $R_M = 1737$ km and the acceleration due to the Moon's gravity is $g_M = 0.00162$ km/s². Its surface escape velocity is:

$$s_e = \sqrt{(2g_M R_M)}$$

$$s_e = \sqrt{[2*(0.00162 \text{ km/s}^2)*(1737 \text{ km})]}$$

$$s_e = \sqrt{(5.63)} \text{ km/s}$$

$$s_e = 2.4 \text{ km/s}$$

This is the same as the gravitational escape velocity at the surface. However, the effect of the gravitational pull of the Earth and Sun on the escape velocity have not been taken into consideration.

Sun

The radius of the Sun is about $R_S = 6.955*10^5$ km and its acceleration due to gravity is $g_S = 0.274$ km/s². The escape velocity from the Sun's gravity is:

$$s_e = \sqrt{(2g_S R_S)}$$

$$s_e = \sqrt{[2*(0.274 \text{ km/s}^2)*(6.955*10^5 \text{ km})]}$$

$$s_e = \sqrt{(38.113*10^4)} \text{ km/s}$$

$$s_e = 617.4 \text{ km/s}$$

Typically, atomic particles escaping the Sun in a solar storm would reach such great altitudes to escape that the $s_e = \sqrt{(2g_S R_S)}$ equation would not be valid.

Problems with equation near Earth

The calculated escape velocity from gravity near the Earth's surface of 11.2 km/s or 25,055 miles per hour is too high to be practical. Also, the effect of the Sun is not taken into account.

Assumes extremely high acceleration

A major problem with the escape velocity from gravity value is that the velocity is calculated at or near the Earth's surface. An infinite acceleration would be required to project an object at 11.2 km/s from the Earth's surface.

Also, it would be very difficult—if not impossible—for a rocket to attain a velocity of 11.2 km/s relatively close to the Earth's surface.

The Saturn rocket that was used to go to the Moon did not reach that speed until it was over 193 km (180 miles) from the Earth's surface.

> (See chapter *13.3 Gravitational Escape Velocity with Saturn V Rocket* for more information.)

Escape velocity value becomes inaccurate

At the 193 km altitude, the gravity escape velocity equation is inaccurate. Instead of being 11.2 km/s, the value for s_e using the gravitation escape velocity equation is 11.007 km/s.

You could get that value from the gravity escape velocity equation, provided you recalculated **g** for that altitude and added 193 km to the radius. However, it would be better to simply use the gravitation escape velocity equation.

Rocket would burn up

Also, at lower altitudes, the rocket would be traveling at hypersonic speed, which would be so far above the speed of sound that it could cause the burn-up of a rocket exterior before it left the Earth's atmosphere.

Realistically, a rocket would build up its speed until it reached the extreme upper atmosphere, where air resistance is negligible at high speeds.

Factor of the Sun

Adding to the escape velocity from the Earth is the gravitational pull from the Sun. This is another factor that must be considered, requiring the gravitational escape velocity equations and not the gravity equations.

> (See chapter *13.4 Effect of Sun on Escape Velocity from Earth* for more information.)

Equation not practical

Although the $s_e = \sqrt{(2gR_E)}$ equation gives a correct value of the escape velocity near Earth, it really is more of an approximation for actual applications.

> **Note**: It pains me to see this value being so carelessly used. It is just unscientific.

Summary

The velocity an object projected upward from the Earth must attain such that it will overcome the gravitational pull and not fall back to the ground can be approximated by the equation $s_e = \sqrt{(2gR_E)}$, provided the object starts near the Earth's surface. A similar equation can be used for the Moon, Sun and other celestial bodies.

However, the equation is not applicable for a rocket blasting off from Earth, due to the extreme acceleration required, effects of the atmosphere and the gravitation from the Sun. Instead, the gravitational escape velocity equation should be used.

Mini-quiz to check your understanding

1. Why is it sometimes called surface escape velocity?

    a. Because it is the attained velocity at or near the surface

    b. To avoid confusion with underwater escape velocity

    c. No one is sure where the expression came from

2. Why is the escape velocity of the Moon so much less than the Earth's?

    a. Because the Moon is so far from the Earth

    b. Because the gravity and radius of the Moon are much less than the Earth's

    c. The escape velocity from the Moon is actually higher than from Earth

3. What might happen if a rocket blasted off at the gravity escape velocity from the Earth?

    a. The rocket exterior might burn up before it left the atmosphere

    b. The rocket would only make it to the Moon

    c. The rocket would go into space and never come back

Answers

1a, 2b, 3a

6.8 Artificial Gravity

Artificial gravity is a force that simulates the effect of gravity but is not caused by the attraction to the Earth. There is a need for artificial gravity in spacecraft to counter the effect of weightlessness on the astronauts.

Acceleration and centrifugal force can duplicate the effects of gravity. Albert Einstein used the concept of artificial or virtual gravity in his *General Theory of Relativity* to give a different explanation of gravity.

A rotating circular space station can create artificial gravity for its passengers. The rate or rotation necessary to duplicate the Earth's gravity depends on the radius of the circle.

Equations can be derived to determine the rotation rate and radius to simulate the effect of gravity.

Where artificial gravity is needed

Artificial gravity is needed in spaceships that are in orbit around the Earth, as well as ones that are so far out that the effect of gravity or gravitation is negligible.

The International Space Station orbits around the Earth at approximately 350 km. Because the centrifugal force keeping the space station in orbit counters the force of gravity at that altitude, astronauts in the station do not feel the effect of gravity.

Anything or anybody that is not tied down will float within the Space Station.

Astronauts in any spaceship that is far enough away from the Earth that the effect of gravity or gravitation is negligible will also feel the effects of weightlessness. The gravitation on a spaceship that is about 15,000 km from Earth is about 1/10 the gravity on the ground.

Thus, artificial gravity is needed to facilitate the tasks the astronauts must do, to make them more comfortable and to avoid negative health effects from weightlessness.

Ways to create artificial gravity

Constant acceleration and centrifugal force are ways to create artificial gravity, such that a person could not tell the force was not gravity and all the laws of gravity hold.

Acceleration

One way to simulate a gravitational force is to accelerate the spaceship. This is similar to the effect you feel when you are in an accelerating elevator, where you can feel heavier when the elevator is moving upward.

In developing his General Theory of Relativity, Albert Einstein noted that you could not tell the difference between gravity and constant acceleration. He used this example to state his theory that gravity or gravitation was not a force but an action related to inertia on moving objects.

Unfortunately, creating artificial gravity is impractical be depending on acceleration alone. There is a limit to the velocity of a spaceship.

Centrifugal force

A better way to create this artificial gravity than constant acceleration is to use centrifugal force, which is an outward virtual force caused by an object being made to follow a curved path instead of a straight line, as dictated by the *Law of Inertia*.

If a spaceship was in a large, circular shape that was rotating at a given speed, the crew on the inside could feel the centrifugal force as artificial gravity.

In the 1968 movie *2001: A Space Odyssey*, a rotating centrifuge in the spacecraft provided artificial gravity for the astronauts. A person could walk inside the circle with his feet toward the exterior and his head toward the center, the floor and ceiling would curve upwards.

A rotating spacecraft will produce the feeling of gravity on its inside hull. The rotation drives any object inside the spacecraft toward the hull, thereby giving the appearance of a gravitational pull directed outward.

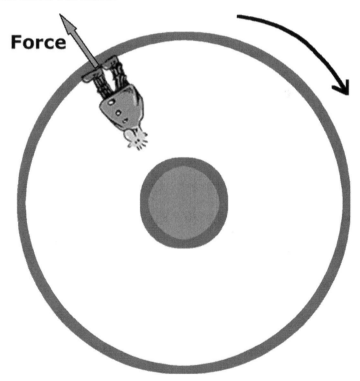

Rotating space station creates artificial gravity

Rate of rotation to duplicate gravity

It is worthwhile to determine the radius of the space station centrifuge and its rate of rotation that will simulate the force of gravity.

Centrifugal force equation

When you swing an object around you that is tied to a string, the outward force is equal to:

F = mv²/r

$$F = mv^2/r$$

where

- **F** is the outward force of the object in newtons (N) or pounds (lb)
- **m** is the mass of the object in kilograms (kg) or pound-mass (lb)
- **v** is the linear or straight-line velocity of the object in meters/second (m/s) or feet/second (ft/s)
- **r** is the radius of the motion or the length of the string in m or ft

Verify units

It is a good practice to verify that the units you are using are correct for the equation.

F N = (m kg)(**v** m/s)²/r m

N = (kg)(m²/s²)/m

kg·m/s² = kg·m/s²

A similar verification can be done using feet and pounds.

Angular velocity equation

A better way to write the force equation is to use angular velocity, which will then lead to revolutions per minute.

ω = v/r

v = ωr

where **ω** (lower-case Greek letter omega) is the angular velocity in radians per second.

> **Note:** A radian is the distance along a curve divided by the radius

Substituting for **v** in $\mathbf{F} = \mathbf{mv^2/r}$, you get

$$\mathbf{F} = \mathbf{m\omega^2 r}$$

Relate to gravity

Since the centrifugal force is $\mathbf{F} = \mathbf{m\omega^2 r}$ and the force due to gravity is $\mathbf{F} = \mathbf{mg}$, you can combine the two equations to get the relationship between the radius, rate of rotation and **g**:

$$\mathbf{mg} = \mathbf{m\omega^2 r}$$

$$\mathbf{g} = \mathbf{\omega^2 r}$$

Solving for **ω**:

$$\mathbf{\omega} = \mathbf{\sqrt{(g/r)}}$$

Also, solving for **r**:

$$\mathbf{r} = \mathbf{g/\omega^2}$$

Convert radians per second to rpm

The units for **ω** are inconvenient for defining the rate of rotation of the space station. Instead of using radians per second, it would be better to state the units as revolutions per minute (rpm). Conversion factors are:

1 radian = 1/2**π** of a full circle (**π** is "pi", which is equal to about 3.14)

ω radians per second is **ω**/2**π** is revolutions per second

ω/2**π** revolutions per second is 60**ω**/2**π** revolutions per minute

60**ω**/2**π** = 9.55**ω** rpm

Thus:

ω radians/second = 9.55**ω** rpm

1 radian/second = 9.55 rpm

Let Ω (capital Greek letter omega) be the rate of rotation in rpm.

$\Omega = 9.55\omega$ rpm

Thus:

$\Omega = 9.55\sqrt{(g/r)}$

and

$r = 91.2g/\Omega^2$

Example 1

Suppose the space station had a radius of $r = 128$ ft. How fast would it have to turn to create an acceleration due to gravity of $g = 32$ ft/s²?

Solution

$\Omega = 9.55\sqrt{(g/r)}$

$\Omega = 9.55\sqrt{(32/128)}$ rpm

$\Omega = 9.55\sqrt{(1/4)}$ rpm

$\Omega = 9.55/2$ rpm

$\Omega = 4.775$ rpm

Example 2

If you wanted the space station to rotate at only 2 rpm, how many meters must the radius be to simulate gravity?

Solution

$r = 91.2g/\Omega^2$

$r = (91.2)(9.8)/(2^2)$ meters

$r = 233.44$ m

Summary

Artificial gravity is a force that simulates Earth's gravity. There is a need for artificial gravity in spacecraft to counter the effect of weightlessness on the astronauts.

Acceleration and centrifugal force can duplicate the effects of gravity. A rotating circular space station can create artificial gravity for its passengers. The rate or rotation necessary to duplicate the Earth's gravity depends on the radius of the circle.

Mini-quiz to check your understanding

1. Where could you feel artificial gravity on Earth?

    a. It only exists in spaceships

    b. When an elevator goes up rapidly

    c. All gravity on Earth is artificial

2. What is a disadvantage of artificial gravity on a rotating space station?

    a. Things fly out of your hand toward the center

    b. You can get dizzy when at the top of the rotation

    c. The floor is curved

3. As the radius of the cylinder increases, what happen to the rotation speed?

    a. The larger the radius, the slower the necessary rotation

    b. The larger the radius, the faster the station must rotate

    c. There is no relationship between the two

Answers

1b, 2c, 3a

6.9 Center of Gravity

The center of gravity (CG) of an object is the balance point around which there are equal moment arms of length times weight. The object can act as if all its weight was concentrated at the CG.

You can find the center of gravity mathematically by taking the average or mean distribution of the weight of the object. You can also find the CG experimentally by either using a plumb line or finding the balance point.

> **Note**: Some textbooks confuse center of gravity with center of mass (CM). Finding the center of gravity requires that the object is under the influence of gravity, while center of mass is the center of a mass distribution.
>
> Although CG is often at the same location as the CM, they are completely different concepts.
>
> (See chapter *11.1 Overview of Gravitation and Center of Mass* for more information.)

Applications include the fact that free rotation of an object is always around its center of gravity and that an object will tip over when the center of gravity lies outside the supporting base of the object. Also, the greatest force is applied through the center of gravity.

Calculation of the center of gravity

Calculation of the center of gravity is based on the fact that a torque exerted by the weight of a system is the same as if it's total weight were located at the center of gravity. This point is the average or mean of the distribution of the weight of the object.

The calculation of the center of gravity of an object involves the summation of the weights times their separations from a starting point divided by the total weight of the object.

In the case of a highly irregular object, the weights can consist of individual particles or even atoms. Calculus is then used to integrate the product of these weights and the differential separations.

If the object is made up of regular parts, such as squares or circles, you can use the fact that each has a CG at its geometric center. This is seen in the illustration below:

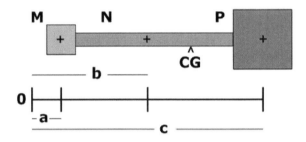

Calculating CG of weights

The center of gravity in the illustration is at the following separation from the arbitrary zero-point:

$$CG = (aM + bN + cP)/(M + N + P)$$

For example, if:
- **a** = 1 ft
- **b** = 4 ft
- **c** = 8 ft
- **M** = 1 lb
- **N** = 2 lb
- **P** = 4 lb

CG = (1*1 + 4*2 + 8*4)/(1 + 2 + 4) ft

CG = 41/7 ft

CG = 5.9 ft from the zero point

The approximate CG is shown in the illustration.

Finding CG through experimentation

The center of gravity for objects with regular shapes, such as squares, cubes, circles and spheres, is at their geometric centers.

For other shapes or configurations, you need to either use experimentation or calculation. It is easier to determine the center of gravity of many objects by experimentation than by calculation.

Plumb line technique

When an object is suspended so that it can move freely, its center of gravity is always directly below the point of suspension.

You can find the center of gravity in an object experimentally by hanging it from several points and using a plumb line to mark the vertical line.

The intersection of two or more vertical lines from the plumb line is the center of gravity for the object.

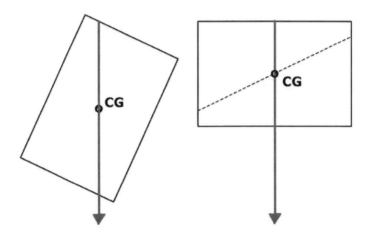

Measuring CG with plumb line

This procedure is relatively easy for a flat object. However, it can be more difficult if the object has some shape in the three dimensions.

Balance point

For some objects, you can find its balance point through experimentation.

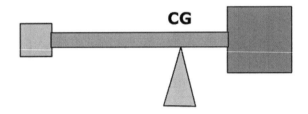

Weight balances on sharp edge

For the object in the illustration above, the balance point could easily be found. However, it would be difficult to find the balance point for an object with curved surfaces.

Applications

A lower center of gravity helps prevent an object—such as an automobile—from tipping over.

Objects will spin about their CG. Also, the center of gravity provides the greatest impact in a collision.

Tipping point

When you tilt an object on an edge, it will tip over only when the center of gravity lies outside the supporting base of the object.

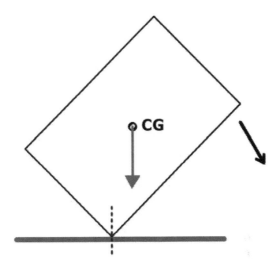

Object tips over when CG passes pivot point

Objects that are heavier toward the bottom have a lower center of gravity and are thus more difficult to tip over. Automobiles and trucks have a lower center of gravity to improve their stability.

CG below balance point

An interesting application is when the center of gravity is below the balance or pivot point. The object will readily balance and may swing back and forth in that position.

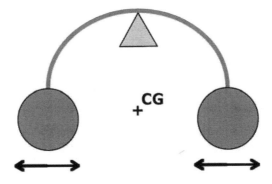

CG below balance point

There are a number of novelty items and toys that use this principle.

Spins about center of gravity

If you throw an object in the air with a spin on it, the object will rotate around its center of gravity as it follows its path.

A ball has its center of gravity at its center, so it will simply spin as it is thrown in the air. The ball will follow a parabolic path until it hits the ground.

Likewise, if you threw a baseball bat in the air, it too would rotate about its center of axis, and that axis would follow a parabolic path similar to that of the ball.

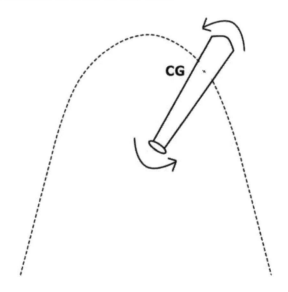

Bat follows parabolic path as it spins

Hitting a baseball

The center of gravity is a point where all of the weight of the object is concentrated.

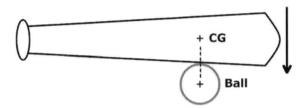

CG is best location to hit the ball

When you swing a baseball bat, the best location to hit the ball is at the bat's center of gravity. At that point, the maximum force is applied to the ball for a given swing.

Summary

The center of gravity (CG) is where all of the weight of an object appears to be concentrated. This point is the average distribution of the weight of the object. The center of gravity of an object can also be found experimentally.

Free rotation of an object is always around its center of gravity. An object will tip over when the CG lies outside the object's support. The greatest force is applied through he center of gravity.

Mini-quiz to check your understanding

1. How do you calculate the center of gravity?

   a. Add up the weights and divide by the separations

   b. Add the product of weights and separations and then divide by the total weight

   c. Measure the total weight, multiply times the total separation and divide by two

2. Where do you hang the plumb line when determining the CG?

   a. It doesn't matter, because it always goes through the CG

   b. Only from the corners

   c. From the point where you are suspending the object

3. What happens when you hit a ball outside the bat's center of gravity?

   a. The ball will travel much further

   b. The ball will not go as far as if hit at the CG

   c. The ball will start spinning rapidly

Answers

1b, 2c, 3b

Part 7: GRAVITATION

Gravitation is a fundamental physical property where objects are attracted to each other. These objects include things as large as stars and suns and as small as subatomic particles. Even light is affected by gravitation.

Part 7 chapter

Part 7 has one chapter:

7.1 Overview of Gravitation

This chapter provides an overview of the definition of gravitation, the law that describes gravitation and the several theories that explain how gravitation occurs.

Gravitation Parts

Other parts of the book concerning Gravitation include

Part 8: Theories of Gravitation

Part 8 explains the major theories about what gravitation is and what causes it.

Part 9: Gravitation Principles

This part provides some basic principles and gravitational phenomena.

Part 10: Gravitation Applications

Applications of the principles and equations of gravitation in the real-world, as well as hypothetical experiments, are explained.

Part 11: Center of Mass

This important concept is explained, along with a number of examples.

Part 12: Gravitation in the Universe

This part gives examples concerning how gravitation affects planets and stars.

Part 13: Escape Velocity

The various aspects of gravitational escape velocity are explained.

7.1 Overview of Gravitation

Gravitation is the force of attraction between objects that is proportional to the product of their masses and inversely proportional to the square of their separation.

Objects can range in size from sub-atomic particles to celestial masses, such as planets, stars and galaxies. Other properties of gravitation include attraction to the center or mass, escape velocity and gravity.

The concept that matter attracts other objects was formulated by Isaac Newton as the *Law of Universal Gravitation*. This theory has been superseded by newer theories of gravitation, such as Albert Einstein's *Theory of General Relativity* and the *Theory of Quantum Gravitation*.

The *Universal Gravitation Equation* defines the force of attraction between two objects in ordinary situations. The equation can be simplified to give the gravity equation for objects near Earth.

Properties of gravitation

All objects consisting of matter exhibit the property of gravitational attraction and tend to move toward each other. This property is considered universal and exists throughout the Universe.

No shield

As far as we know, there is no way to shield the effect of gravitation. There are theories that there exists "dark matter" that repels standard matter, however dark matter has never been detected.

Center of mass

There is a center of mass between disconnected objects, which is an average of the masses and their separations. When the objects move toward each other, the will meet at the center of mass.

If one object seems to be revolving around another, as in the case of a moon around a planet, both objects are actually revolving around the center of mass.

Escape velocity

It is possible for an object to be propelled at a sufficient velocity away from another object that it will overcome the gravitational attraction between the two. An example of this is when a rocket escapes the gravitation of the Earth.

Gravity

The expressions *gravity* and *gravitation* are often commonly interchanged. However, the correct scientific terminology considers gravity as a special case of gravitation for objects near the Earth.

> For gravitation close to other large objects, you should include the name of the object, such as: "gravity of the Moon" or "gravity of the Sun."

For astronomical situations, gravitation is the correct term to use.

Gravitational theories

There have been several theories trying to explain the cause of gravitation.

Law of Universal Gravitation

In 1687, Isaac Newton formulated the *Law of Universal Gravitation*, which states that all objects are attracted toward other objects, due to some force acting at a distance, called gravitation.

The law and its equation apply to moderately-sized astronomical effects and is still the mainstay theory today.

Theory of General Relativity

In 1915, Albert Einstein gave another interpretation of gravitation in his *Theory of General Relativity.*

He stated that gravitation was the result of the curvature of space toward matter and not due to some force. General relativity applies best to very large gravitational fields and high velocities.

Anomalies in the orbit of the planet Mercury, which is closest to the Sun, could not be explained by the Law of Universal Gravitation. However, calculations using General Relativity equations correctly predicted the orbit and were used to verify the theory, as were the measurements of the effect of gravitation on deflecting light waves as they pass a star.

Theory of Quantum Gravitation

Recent considerations in *Quantum Physics* say that gravitation is one of four fundamental forces in nature.

The force of each is created by an exchange of special or virtual particles. In the case of gravitation, the particle is called the graviton. This interaction leads to an explanation of gravitation at very small separations.

Dark matter and dark energy

Astronomical measurements on the rate of expansion of the Universe and measurements of the mass of galaxies showed discrepancies with predictions from general relativity.

To explain the discrepancies, astronomers have speculated that there exists a form of dark matter and dark energy that they cannot detect.

These concepts fit to a degree in the Theory of Quantum Gravity.

Universal Gravitation Equation

Newton formulated the *Universal Gravitation Equation,* which allows the calculation of the force between two objects. The equation is:

$$F = GMm/R^2$$

where

- **F** is the force of attraction between two objects in newtons (N)
- **G** is the universal gravitational constant in N-m^2/kg^2
- **M** and **m** are the masses of the two objects in kilograms (kg)
- **R** is the separation in meters (m) between the objects, as measured from their centers of mass

Gravity equation

The gravity equation is a simplification of the gravitational equation for objects relatively close to the Earth:

$$F = mg$$

where

- **F** is the force pulling objects toward the Earth in newtons (N) or pound-force (lbs)
- **m** is the mass of the object in kg or pound-mass
- **g** is the acceleration due to gravity in meters per second squared (m/s^2) or feet per second squared (ft/s^2)

Other equations

The Theory of General Relativity provides a set of 10 complex equations to describe gravitation.

Likewise, Quantum Mechanics and other newer theories explain gravitational force with sophisticated equations.

They are beyond the scope of our material.

Summary

Gravitation is the attraction between objects because of their mass. Other properties of gravitation include attraction to the center or mass, escape velocity and gravity.

Theories of gravitation are the Law of Universal Gravitation, the Theory of General Relativity and the Theory of Quantum Gravity.

The Universal Gravitation Equation defines the force of attraction between two objects in ordinary situations. The equation can be simplified to give the gravity equation for objects near Earth.

Mini-quiz to check your understanding

1. What is the difference between gravitation and gravity?

    a. They are different spellings of the same word

    b. Gravitation attracts and gravity repels

    c. Gravity is gravitation on or near the Earth

2. What was one proof that gravitation is due to the curvature of space?

    a. Light passing by huge masses such as the Sun was deflected

    b. No one has figured out how to prove it yet

    c. A ball thrown in the air follows a curved path

3. How does the force of gravitation vary with separation?

    a. It gets stronger the further you are from an object

    b. It is independent of distance

    c. It decreases the further you are from an object

Answers

1c, 2a, 3c

Part 8: Theories of Gravitation

There are several theories about what gravitation is and what causes it. This section covers the major theories.

Part 8 chapters

Chapters in Part 8 include:

8.1 Overview of the Theories of Gravitation

This chapter gives an overview of Isaac Newton's concept of gravitation and Albert Einstein's General Relativity Theory, as well as the Quantum Theory of Gravitation.

8.2 Law of Universal Gravitation

Isaac Newton established a law concerning the gravitational attraction of all objects in the Universe and gave an equation for that law. However, there were some problems with Universal Gravitation.

8.3 Universal Gravitation Equation

This chapter explains the Universal Gravitation Equation and its Gravitational Constant, as well as shows how the equation an approximation.

8.4 General Relativity Theory of Gravitation

This chapter explains Einstein's theory that gravitation is caused by the curvature of space, gives predictions proving the concept and includes problems with the relativity explanation.

8.5 Theory of Quantum Gravitation

Quantum Gravitation is an effort to explain gravitation on the atomic and subatomic levels. There are several Quantum Gravitation theories and problems concerning them.?

8.6 Effect of Dark Matter and Dark Energy on Gravitation

Dark matter seems to add to the gravitation caused by the visible matter. Dark energy seems to be a force that acts opposite of gravitation. This chapter explains these strange phenomena, as well as problems with the concepts.

8.7 Gravitation as a Fundamental Force

This chapter shows how gravitation can be considered a fundamental force, along with nuclear forces and electromagnetic forces.

8.1 Overview of Theories of Gravitation

Since ancient times, scientists used observations about the effect of gravity on objects near the Earth (like the proverbial apple falling from a tree), as well as measurements of the movement of planets, to establish laws about the properties of gravitation and theories about possible causes of the phenomenon.

The first major concept was Newton's *Law of Universal Gravitation*, which stated that all objects of matter attract each other and provided an equation to measure the force of attraction.

Albert Einstein reformulated the gravitation laws to fit his *Theory of General Relativity*, explaining that gravitation is caused by a curvature of space and especially applies to large gravitational fields.

Theories in *Quantum Mechanics* about gravitation connect it to the fundamental forces of matter and state that gravitation has wavelength and is a particle.

Law of Universal Gravitation

Isaac Newton combined concepts from other scientists to define the *Law of Universal Gravitation* in 1687. This scientific law states that all objects consisting of matter are attracted toward other objects, due to a force called gravitation.

Universal gravitation assumes that the law holds throughout the Universe.

> **Note:** Newton created a law, as opposed to a theory. A law is based on observations and measurements but does not explain the reasons for something happening, as does a theory.

Newton also established the *Universal Gravitation Equation*, which can be used to calculate the force of attraction between two objects, provided you know their respective masses and the separation between their centers:

F = GMm/R²

This equation allowed the calculation of orbits of the planets, as well as other motion problems.

(See chapter *8.2 Law of Universal Gravitation* for more information.)

This classical view of gravitation is that objects attract each other through some force that acts at a distance. It was uncertain what the mechanism was for that force and how it tied into other forces that acted at a distance.

Although measurements could be made, the fact there was not a good explanation for gravitation differentiated the law from being a theory.

Theory of General Relativity

Although the Law of Universal Gravitation was able to predict many phenomena, there were some experimental measurements—notably the variations in the orbit of the planet Mercury around the Sun—where it was not accurate.

(See chapter *8.4 General Relativity Theory of Gravitation* for more information.)

This problem helped initiate Albert Einstein's *Theory of General Relativity* in 1915. This theory explained gravitation, especially in areas of high intensity fields, such as with a planet like Mercury that is relatively close to the Sun.

Curvature of space

Einstein postulated that the presence of matter changes the geometry of space, such that it curves the space around it. The straight line that at freely moving object traveled would curve toward another object of mass, resulting in the effect of gravitation.

In effect, Einstein claimed that gravitation was not really a force but simply a result of this property of space being curved. This is a pretty abstract concept, but Einstein demonstrated it with highly complex mathematical equations.

Experimental proof

A number of experimental measurements proved very accurate using Einstein's equations, including the orbit of Mercury, the bending of light by a large gravitational field and properties of Black Holes.

This brought on the acceptance of the Theory of General Relativity as an explanation of gravitation.

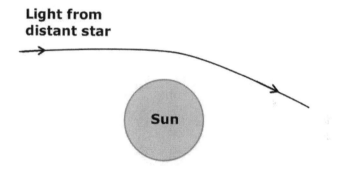

Sun's gravitation bends beam of light

Quantum theories

Although the Theory of General Relativity worked fine on large scales, it seemed to fail at the subatomic level and did not follow the rules of *Quantum Mechanics* or *Quantum Physics*—the study of matter at extremely small separations.

(See chapter *8.5 Quantum Theory of Gravitation* for more information.)

One problem with the Theory of General Relativity's explanation of gravitation as being caused by the curvature of space is the question why gravitation is so different than other fundamental forces that act at a distance—such as magnetism.

Gravitation as a waveform

Scientists then predicted that a gravitational field would exhibit wavelengths, similar to electromagnetic waves. Using highly sensitive instruments, experiments were made to verify that gravity is indeed a waveform of some sort.

Wave-particle duality

To follow the wave-particle duality in Quantum Physics, it was also predicted that gravity consists of particles called *gravitons*.

This is similar to the theory that light is not only a waveform, but also consists of particles called photons. Electrons are also viewed as both particles and waves.

This view of gravity being a wave or particle goes back to the idea that matter exhibits gravitation and that there truly is a force of gravity. (Nobody said that Physics would be easy and that everything was known or explained.)

Quantum Mechanics

Recently there have been new theories that the force of gravity is caused by graviton particles or by gravity waves. These theories satisfy rules of Quantum Mechanics that Einstein's concepts didn't.

Dark matter

There is also a theory that there exists some sort of "dark matter" that repels instead of attracts, resulting in anti-gravitation.

(See chapter *8.6 Effect of Dark Matter and Dark Energy on Gravitation* for more information.)

Summary

Newton's *Law of Universal Gravitation* said that all objects of matter attract each other. He also provided an equation to measure the force of attraction.

Over 200 years later, Albert Einstein reformulated the gravitation laws to fit his *Theory of General Relativity*, stating that gravitation is caused by a curvature of space. *Quantum Mechanics* theories about gravitation connect it to the fundamental forces of matter and state that gravitation has wavelength and is a particle.

Mini-quiz to check your understanding

1. Why did Newton say that gravitation is a force at a distance?

    a. Newton could not think of a better name

    b. The force acts even when there is no physical contact between objects

    c. The force disappears at a distance

2. According to Einstein's theory, what happens as you increase the mass of an object?

    a. Space curves more towards the object

    b. The object falls faster

    c. It becomes relative

3. How can waves be particles at the same time?

    a. It is part of the wave-particle duality theory

    b. It is impossible

    c. It happens when they fall in an elevator

Answers

1b, 2a, 3a

8.2 Law of Universal Gravitation

In 1687, Isaac Newton combined his observations with theories from other scientists about gravity and gravitation into a scientific law: the *Law of Universal Gravitation*. This law states that the mass of an object is attracted toward the mass of other objects by a force called gravitation.

The force of attraction between two masses is defined by the Universal Gravitation Equation. Although the law and its equation were effective in predicting many phenomena, several discrepancies later arose in astronomical measurements. It wasn't until 1915 that Einstein's *Theory of General Relativity* provided a solution to these discrepancies.

Establishing Law of Universal Gravitation

Scientists and philosophers from ancient times have made observations about gravity on Earth. In 628, Indian astronomer Brahmagupta recognized gravity as a force of attraction. In the 1600s, Galileo Galilei, Robert Hooke and Johannes Kepler formulated laws of gravity near the Earth.

Newton plays off Kepler's Laws

In studying the orbits of planets around the Sun, Kepler determined that gravitational attraction varied with separation, while formulating his *Laws of Planetary Motion*.

In 1687, Isaac Newton's observations on planetary motion and empirical measurements, allowed him to establish the *Law of Universal Gravitation*, which was explained in the publication of *Philosophiæ Naturalis Principia Mathematica* (or simply *Principia*) set of three books that stated his Laws of Motion, the Law of Universal Gravitation and a derivation of Kepler's Laws of Planetary Motion.

> The story that Newton was sitting under a tree when he was hit in the head by a falling apple, thus causing him to discover gravity, is a fable.

> Although he may have been inspired by seeing an apple fall from a tree, his studies and correspondence with other scientists who were studying gravitation probably influenced him more than seeing an apple fall. Nevertheless, it is a good story.

Statement of Universal Gravitation

The Law of Universal Gravitation states that every object of mass in the Universe attracts every other object of mass with a force which is directly proportional to the product of their masses and inversely proportional to the square of the separation between their centers.

This was then formalized into the *Universal Gravitation Equation.*

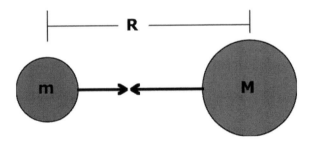

Masses attract each other

This law was originally stated for point masses. However, it was shown that the gravitation from a large uniform sphere is approximately the same as if all the mass was concentrated at its center.

Law versus theory

Note: Newton's *Law of Universal Gravitation* is sometimes incorrectly called the *Theory of Universal Gravitation*. In the sciences, a law generalizes observations and shows no exceptions. On the other hand, a theory tells why a phenomenon happens. The Law of Universal Gravitation only makes predictions of how bodies are attracted and does not explain why it happens.

Universal Gravitation versus Universal Law

Note: Also, the *Law of Universal Gravitation* is sometimes called the *Universal Law of Gravitation*. A universal law is one that is said to apply to the whole Universe. On the other hand, universal gravitation implies that gravitation is throughout the universe, which is slightly different than a universal law.

Force at a distance

Newton could not explain the mechanism of how and why gravitation occurred, except that it was some sort of *force at a distance*. Critics said that this explanation was bringing the occult or mysticism into science, which Newton denied.

Universal Gravitation Equation

From the Law of Universal Gravitation, Newton formulated the *Universal Gravitation Equation*, which defines the gravitational force between two objects. The equation is:

$$F = GMm/R^2$$

where

- **F** is the force of attraction between two objects in newtons (N)
- **G** is the Universal Gravitational Constant = $6.67 * 10^{-11}$ N-m^2/kg^2
- **M** and **m** are the masses of the two objects in kilograms (kg)
- **R** is the separation in meters (m) between the objects, as measured from their centers of mass

---

Useful in determining orbits

Application of this equation is especially useful in determining the orbits of celestial objects, such as that of the planets around the Sun. This is relatively simple when considering one object orbiting another.

(See chapter *12.4 Circular Planetary Orbits* for more information.).

However, in most cases there are more than two objects involved, which can result in some complicated mathematics.

For example, the motion of the planet Mars around the Sun is not only determined by the masses of Mars and the Sun but also by the influence of the gravitation from Mar's moons, the Earth and other planets. The result is that orbit of Mars includes perturbations or deviations caused by the other planets.

Problems with Universal Gravitation

In most cases, the Law of Universal Gravitation and resulting calculations work well. But there are some observed discrepancies and unexplained phenomena where the law fails.

Orbit of planet Mercury

Predictions of the orbit of the planet Mercury proved to be slightly inaccurate using the Law of Universal Gravitation. The law could not explain the precession or deviations of Mercury's orbit.

Deflection of light rays

Electromagnetic waves or light rays are deflected by gravitation, especially when they pass by a large mass, such as the Sun. However, calculations from Newton's law result in only one-half of the deflection that is actually observed by astronomers.

Speed of gravitation

The Law of Universal Gravitation requires that the gravitational force is transmitted instantaneously over large distances to maintain stability in planetary and stellar orbits. Although light from the Sun may take eight minutes to reach the Earth, gravitational changes would have to occur immediately to maintain orbital stability between the two objects.

This implies that what is called the *speed of gravitation* (also carelessly called the *speed of gravity*) is infinite. That seems counterintuitive.

Solutions found

Solutions to these problems were finally found in Einstein's *Theory of General Relativity* that he published in 1915. The theory reduces to the Law of Universal Gravitation for simple cases..

Summary

Since ancient times, scientists and philosophers studied gravity and then later gravitation.

Isaac Newton formalized the observations into a scientific law: the *Law of Universal Gravitation*, which states that states that all objects of mass are attracted toward other objects of mass, due to a force called gravitation.

The force of attraction between two masses is defined by the *Universal Gravitation Equation*. Although the law and its equation were effective in predicting many phenomena, several discrepancies arose in measurements that were made. Einstein's *Theory of General Relativity* provided a solution to the Universal Gravitation problems.

Mini-quiz to check your understanding

1. Did Isaac Newton discover gravity?

    a. Yes, he made the discovery after keen observations

    b. Most people think he invented gravity, because it didn't exist before him

    c. No, Newton elaborated on previous work of other scientists

2. Why would the orbit of Mars be affected by the Earth?

    a. Gravitation from the Earth can distort the orbit of Mars at certain positions

    b. Mars will sometimes rotate around the Earth

    c. The planets are too far apart to have any influence

3. Are light rays deflected by the Earth's gravitation?

    a. No, because light has no weight

    b. Yes, although the deflection would be extremely small

    c. It depends on the speed of the light rays

Answers

1c, 2a, 3b

8.3 Universal Gravitation Equation

The *Universal Gravitation Equation* states that the gravitational force between two objects is proportional to the product of their masses and inversely proportional to the square of separation between them. This equation is a result of Isaac Newton's *Law of Universal Gravitation*, which states that quantities of matter attract other matter to it.

The proportionality constant in the equation is called the *Universal Gravitational Constant*. The value of that constant was determined experimentally by Henry Cavendish in 1798.

This Universal Gravitation Equation originally applied to point masses but was extended to masses of finite size with the assumption that their mass was concentrated at their center of mass.

Universal Equation

In 1687, Isaac Newton originally formulated the *Universal Gravitation Equation*, which defines the gravitational force between two objects. The equation is:

$$F = GMm/R^2$$

where

- **F** is the force of attraction between two objects in newtons (N)
- **G** is the Universal Gravitational Constant in N-m^2/kg^2
- **M** and **m** are the masses of the two objects in kilograms (kg)

- **R** is the separation in meters (m) between the objects, as measured from their centers of mass

This equation has proven highly effective in explaining the forces between objects, as well as leading into the effects of gravity.

Universal Gravitational Constant

When Newton stated the equation, he simply said that **F** was proportional to **Mm/R²**. The value of the proportionality constant or *Universal Gravitational Constant*, **G**, was not even considered for many years and not officially calculated until 1873, 186 years after Newton defined the equation.

The *Cavendish Experiment* has since been used to determine Universal Gravitational Constant as:

$$G = 6.674*10^{-11} \, \text{N-m}^2/\text{kg}^2$$

Note: The number 10^{-11} is $1/10^{11}$ or 0.000000000001, having 11 zeros after the decimal point.

(See chapter *10.2 Cavendish Experiment to Measure Gravitational Constant* for more information.)

Check on units

It is important to make sure you are using the correct units for each item in your equation. Check by adding units to the gravitation equation and then seeing that the result is correct:

$$\textbf{F N} = \textbf{(G N-m}^2/\textbf{kg}^2\textbf{)(M kg)(m kg)/(R m)}^2$$

Just considering the units:

$$N = (\text{N-m}^2/\text{kg}^2)*(\text{kg})*(\text{kg})/(\text{m})^2$$

$$N = (N)*(\text{m}^2)*(\text{kg})*(\text{kg})/(\text{m}^2)*(\text{kg}^2)$$

$$N = N$$

Thus, the units used are correct.

Other units

G can also be stated in other terms, depending on its usage. Since a newton (N) equals kg-m/s^2, you may also see **G** defined as:

$$G = 6.674*10^{-11} \text{ m}^3/\text{kg-s}^2$$

Also, in applications where greater separations are studied, it is more convenient to use kilometers instead of meters. Since 1 m = 10^{-3} km, the value of **G** is::

$$G = 6.674*10^{-20} \text{ km}^3/\text{kg-s}^2$$

When comparing a force in newtons with gravitational force with km, the value is the strange combination of units:

$$G = 6.674*10^{-17} \text{ N-km}^2/\text{kg}^2$$

You can use whichever set of units that fulfill your requirements.

Equation an approximation

Newton originally stated the Universal Gravitation Equation as the force between two point masses, separated by **R**.

Problem with point masses

A problem exists when considering point masses, and that is the situation when the separation between point masses approaches zero. In such a case, the gravitational force becomes approaches infinity. Since that is impossible, there must be some small separation at which the Universal Gravitation Equation breaks down, perhaps at quantum distances.

Objects of finite size

The concept was later extended to objects of finite size, using the assumption that the mass was concentrated at the center of mass of each object.

In a system of particles, the center of mass is the average of the particle positions, weighted by their masses. The center of mass of a sphere that has its mass evenly distributed is the center of the sphere.

(See chapter *11.1 Overview of Gravitation and Center of Mass* for more information.)

Thus, the separation **R** in the Universal Gravitation Equation is the separation between the objects, as measured from their centers of mass.

Summation of forces

The true gravitational force between two objects is a summation of the forces from each point on both objects.

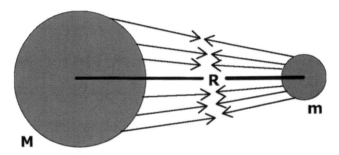

Various points on object attract points on other object

Calculus is used to integrate over all the points on the surfaces and within each object. Unfortunately, the mathematics for the exact equation is highly complex, and it is easier to make some assumptions to simplify the math.

By considering the mass of the objects concentrated is their center of mass, we get an equation that is close enough for practical purposes in most cases.

Distribution of matter in spheres

Most of the objects where the Universal Gravitation Equation apply are large spheres, such as planets, moons and stars. Often the distribution of mass in those objects is not even, and the objects are often not exact spheres.

For example, the density of matter in the Earth is unevenly distributed, plus the Earth is not an exact sphere but is flattened near its poles.

Since the separation between astronomical objects—such as the Earth and the Moon or Sun—are so large, assuming the center of mass as the center of the object is an acceptable approximation.

Consider atoms as points

Atoms, molecules and even subatomic particles are considered so small and separated by great distances relative to their size that they can be considered point sources of gravitation, and the Universal Gravitation Equation applies to these small particles.

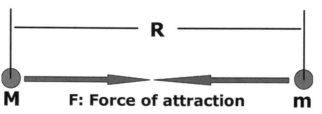

Atoms are considered as points separated by a distance

However, since molecules and atoms are normally in rapid motion, you would seldom calculate the gravitational force between them, except perhaps as an average.

Summary

Isaac Newton formulated the Law of Universal Gravitation, stating that all matter attracts other matter to it. This force of attraction is defined in the theory's Universal Gravitation Equation.

This equation is actually a close approximation, to simplify the mathematics. The measurement of the gravitational constant was first made by Henry Cavendish.

Mini-quiz to check your understanding

1. What was Newton's original equation?

    a. $F = Mm/R^2$

    b. $G = Mm/R^2$

    c. F is proportional to Mm/R^2

2. Why was the Cavendish Experiment important?

    a. It determined the value of G

    b. It disproved Newton's equation

    c. It showed that G varies with the mass and separation

3. What is an assumption for the force between large objects?

    a. Both objects are the same size

    b. The mass of each is considered concentrated at is center of mass

    c. The force is infinite

Answers

1c, 2a, 3b

8.4 General Relativity Theory of Gravitation

In 1915, Albert Einstein formulated the *Theory of General Relativity* as an extension to his *Theory of Special Relativity* and as a new way to explain gravitation.

Newton's *Law of Universal Gravitation* defined gravitation as a property of matter that is force of attraction acting at a distance. The theory works well for ordinary gravitational fields but is inaccurate when the gravitational intensity is high. Discrepancies were seen when measuring the orbit of the planet Mercury and the effect of gravitation on light.

Einstein's theory states that matter curves space and distorts time, causing objects to move toward each other. The Theory of General Relativity became accepted after it predicted the orientation of Mercury's orbit. However, some scientists see flaws in the Theory of General Relativity, due to phenomena that the theory does not explain, especially at the quantum level.

Einstein's theory of gravitation

In 1905, Albert Einstein published the *Theory of Special Relativity*, a theory about space and time. In the following years, Einstein had noticed that acceleration produced the same effect as gravitation.

For example, if you were in an accelerating spaceship—or even an elevator—you could not tell if the force on you was from inertia or gravitation. This led him to look at the mathematics of relative motion and gravitation.

(See chapter *6.8 Artificial Gravity* for more on gravitation from acceleration.)

Then in 1915, he unified the Theory of Special Relativity with Newton's Law of Universal Gravitation to establish the *Theory of General Relativity*, which is a geometric explanation of gravitation.

Curvature of spacetime

While the classical explanation of gravitation is that it is some sort of force-at-a-distance, Einstein took a different approach, stating that matter curves space and distorts time, causing objects to move toward each other. His theory states that time is a fourth dimension, adding to the three dimensions of space. He called this four-dimensional geometry *spacetime*.

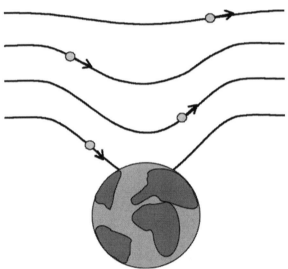

Straight lines curve toward mass

In ordinary situations, particles travel along straight lines in this geometry of space, following Newton's laws of motion. However, according to Einstein's theory, matter affects the geometry of spacetime causing it to be curved toward the matter. A particle that is freely moving at a constant velocity will follow such a line toward the object, as if it was attracted by some force.

Speed of gravitation

Newton's *Law of Universal Gravitation* considered the effects of gravitation to act instantaneously—even over great separations. This meant that the speed of gravitational changes (or *speed of gravitation*) is infinite.

Einstein's equations state that gravitational changes are transmitted at the speed of light, which is the maximum speed for transmitting energy or information. These equations also imply that gravitation can be transmitted as a waveform.

(See chapter *9.3 Gravitational Speed* for more information.)

Mathematics complex

Mathematical expression describing the properties of a gravitational field surrounding a given mass is stated in a set of formulas called the *Einstein Field Equations*. They are highly complex a system of partial differential equations, which are beyond the scope of our material. However, the equations reduce to Newton's Universal Gravitation Equation under simple conditions.

Tests verifying Relativity explanation

The Theory of General Relativity gained acceptance in the scientific community after a several predictions and tests proved correct, including an accurate prediction of the orbit of Mercury and of the deflection of light by a strong gravitational field. However, verification of the theory still remains difficult.

Orbit of planet Mercury

The orbit of the planet Mercury—the closest planet to the Sun— exhibited perturbations and a precession that could not be fully explained by the Law of Universal Gravitation.

Application of the General Relativity equations predicted the motion of Mercury to a high degree. This proof caused most scientists to accept the General Relativity's explanation of gravitation.

Deflection of light

Both Universal Gravitation and General Relativity predict that light can be deflected by gravitation. However the calculation of the amount of deflection from Newton's theory was only half of what Einstein predicted.

Several years after the General Theory of Relativity was proposed, scientists measured the deflection of light from a star as it passed by the Sun during a solar eclipse. Measurements agreed with Einstein's predictions.

Other predictions

The General Theory of Relativity also predicted light coming from a strong gravitational field would have its wavelength shifted toward longer wavelengths, called a *red-shift*. The theory also predicted the existence of Black Holes.

> (See chapter *13.5 Gravitational Escape Velocity from a Black Hole* for more information.)

Both gravitational red-shift and Black Holes were also considered possible in the Universal Gravitation theory, but measurements corresponded better with Einstein's theory.

Problems with General Relativity theory

Although General Relativity does a good job of explaining gravitation at very high levels, it does run into some problems indicating it may not be a complete theory.

Gravitation and other forces

Gravitation, nuclear, magnetism and electrical forces are fundamental entities in the classical, as well as Quantum Mechanics theories.

However, General Relativity does not look at gravitation as a force but instead a property of spacetime. This discrepancy between gravitation and the various forces is a concern to some scientists that the relativity theory is not complete.

(See chapter *8.7 Gravitation as a Fundamental Force* for more information.)

Dark matter

Measurements on the rate of expansion of the Universe indicate there is some force slowing it down. Some scientists proposed there exist what they called *dark matter*, which seemed to have the property of *anti-gravitation*. The possibility of an anti-gravitation force is not explained in the Theory of General Relativity.

(See chapter *8.6 Effect of Dark Matter and Dark Energy on Gravitation* for more information.)

Hofava theory gives different view

A very recent theory called *Hofava gravitation* discounts the spacetime connection and states the space and time are separate entities. The Hofava concept looks at gravitation during the Big Bang and works well with Quantum Gravitation.

This theory has had good results in predicting gravitational phenomenon, but it is new and does not yet have universal acceptance from the scientific community.

Summary

Albert Einstein formulated the *Theory of General Relativity* as a new way to explain gravitation. The theory states that matter curves space and distorts time, causing objects to move toward each other.

While Newton's *Law of Universal Gravitation* works well for ordinary gravitational fields, it is inaccurate when the gravitational intensity is high.

The Theory of General Relativity became accepted by scientists after it correctly predicted the planet Mercury's orbit. However, the theory has areas it does not explain, such as at the quantum level.

Mini-quiz to check your understanding

1. What is spacetime?

    a. The combination of the three dimensions of space with time

    b. The time is takes an object to travel in space

    c. Gravitation

2. Why would the planet Mercury have an unusual orbit?

    a. Mercury is in the realm of spacetime, thus causing relativity from the Sun

    b. Mercury is a shiny liquid planet, thus affecting its motion

    c. It is so close to the Sun that the high gravitational field affects its motion differently than planets further away from the Sun

3. What is one difficulty with the General Relativity explanation?

    a. Gravitation doesn't fit with other fundamental forces

    b. No one, except Einstein, can understand it

    c. Space is empty, so it cannot be curved

Answers

1a, 2c, 3a

8.5 Quantum Theory of Gravitation

Quantum Gravitation (or Quantum Gravity) is an effort to explain gravitation on the atomic and subatomic levels.

While the Theory of General Relativity explains gravitation for large scale events and the Law of Universal Gravitation provides an equation for ordinary situations, neither are able to explain gravitation for extremely small sizes and masses where quantum effects take place.

This type of gravitation is based on Quantum Mechanics, which provides a different perspective of matter, energy and space at very small separations.

There are several theories that attempt to resolve how gravitation works at the quantum level, as well as to fit within the General Relativity concepts.

However, there are also some problems concerning quantum gravitation. This is due to the wide range of quantum theories and difficulties proving them for gravitation.

Quantum Mechanics

At the atomic and subatomic levels, classical or Newtonian physics loses its validity and Quantum Mechanics (or Quantum Physics) takes over. There are several main concepts that provide changes.

Energy is discrete

Quantum theory states that physical interactions and exchange of energy cannot be made arbitrarily small. Energy comes in tiny packets called *quanta* (plural of quantum). This helps to

---

explain the properties of electrons in atoms and the relationship of energy to matter.

Fundamental forces

Another Quantum Mechanics concept is that there are three fundamental forces or interactions between particles of mass. The forces occur through the exchange of virtual particles or particles without mass. Each interaction consists of a quantum packet.

Strong nuclear force comes from the exchange of *gluon* particles between nuclear particles, creating the force of attraction. Weak nuclear force comes from the exchange of *vector bosons*. Electromagnetic force is caused by the exchange of virtual *photon* particles.

Wave-Particle duality

A third concept that relate to quantum gravitation is that both light and matter exhibit a combination of wave and particle behaviors.

For example, subatomic particles can appear as waves. An electron can have properties of a waveform. Likewise, waves can appear as particles. An electromagnetic wave can be a photon light particle.

Theories of quantum gravitation

Extending the Quantum Mechanics theories to gravitation is an effort to explain the force at the quantum level. Some concepts try to reconcile quantum gravitation with general relativity.

Gravitation as fourth fundamental force

One aspect of the *Quantum Theory of Gravitation* is that gravitation is the fourth fundamental force. This would unify all the forces or interactions between particles of matter under one concept.

In such a case, gravitation is the weakest of the fundamental forces, with a strength of only $6*10^{-39}$ of the strength of the strong nuclear force at a sub-nuclear separation.

However, at separations beyond the sub-nuclear range, the strength of gravitation is much larger than the nuclear forces, which are essentially zero at those separations. But still, gravitation is only $8.22*10^{-37}$ of the electromagnetic force at the same separations.

Graviton particles

The Quantum Theory of Gravitation provides an explanation of the mechanism of gravitation that is different from the *Law of Universal Gravitation* and *General Relativity Theory of Gravitation.*

It states that gravitation is caused by an exchange of *graviton* particles or quanta between objects.

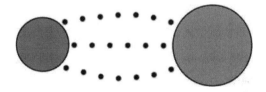

Transfer of gravitons between two molecules

Also, due to the wave-particle duality, this means that gravitation waves are possible.

String Theory

String Theory is an effort to reconcile Quantum Mechanics and General Relativity into a Quantum Theory of Gravitation.

The theory states that particles of matter are one-dimensional oscillating lines or strings. The mathematics of String Theory describes the fundamental forces into a complete system.

One aspect of String Theory is that up to 12 dimensions are required to describe matter and its interactions.

Loop Quantum Gravity Theory

Loop Quantum Gravity also attempts to reconcile the theories of Quantum Mechanics and General Relativity by quantizing the gravitational field. The theory suggests that space consists of moveable tiny loops that can be viewed as an extremely fine fabric.

However, the theory keeps gravitation separate from other fundamental force fields. It also incorporates General Relativity without requiring String Theory's higher dimensions.

Problems with Quantum Gravitation

There are several problems concerning the theories of Quantum Gravitation.

Not universally accepted

One problem is that there are so many theories of Quantum Mechanics that none is universally accepted. This, of course, affects the Quantum Theory of Gravitation.

Does not correspond with General Relativity

Another problem is that Quantum Gravitation does not correspond very well with General Relativity. String Theory and Loop Quantum Gravity Theory make an attempt to find a common ground with General Relativity.

However, a major problem with String Theory is that it may be impossible to prove or even disprove. The same may be true of Loop Quantum Gravity Theory.

Verification extremely difficult

Experimental verification of Quantum Gravitation is extremely difficult, primarily due to the small sizes and weak interactions. Thus far, the existence of gravitons or gravitational waves has never been verified.

Summary

The Quantum Theory of Gravitation is an effort to explain gravitation on the atomic and subatomic levels. It is based on Quantum Mechanics, which provides a different view of matter, energy and space at very small separations.

There are several theories that attempt to resolve how gravitation works at the quantum level, as well as to fit with the General Relativity concepts. However, there are problems in verifying quantum gravitation due to the small forces involved.

Mini-quiz to check your understanding

1. Is a force continuous over extremely small separations?

    a. Yes, as long as you keep applying the force

    b. It depends on the direction of the force

    c. No, the force becomes quantized

2. What is the reason to include gravitation as a fundamental force?

    a. It unifies all the interactions between particles of matter into one concept

    b. It is to avoid arguments between astronomers and physicists

    c. It is a way to explain dark energy and anti-gravitation

3. What is a major problem with the Quantum Theory of Gravitation?

    a. It does not explain how strong gravity is near the Earth

    b. It has been so difficult to verify through experiments

    c. No one really believes it

Answers

1c, 2a, 3b

8.6 Effect of Dark Matter and Dark Energy on Gravitation

Although there are several laws and theories that seem to explain gravitation, astronomical studies have indicated that some other factors may be influencing or may be a part of gravitation. These factors have been called *dark matter* and *dark energy*.

Dark matter is invisible material that seems to add to the gravitation caused by the visible matter in or around galaxies.

The effect was first observed by astronomer Fritz Zwiky in 1933, when making luminosity and Doppler shift measurements to determine galaxy mass. The measurements implied there was more mass available than could be seen, thus causing increased gravitation.

On the other hand, dark energy seems to be a force that acts opposite of gravitation, pushing stars and galaxies apart instead of toward each other. It seems to be a form of anti-gravitation.

Neither dark matter nor dark energy has been directly observed, so each is really a way to explain anomalies in gravitation for objects at the galaxy scale of measurement.

Dark matter

The existence of invisible dark matter can explain why observations on gravitational forces in distant galaxies indicate that there must be more matter than expected.

Background

Upon studying the motion of distant galaxies, astronomer Fritz Zwiky performed measurements to determine their mass by measuring the amount of light they emitted. He then used the

Doppler Effect to find the velocity of the suns moving around various galaxies as another way to determine galaxy mass.

What Zwiky discovered was that the velocity of outer suns were traveling much faster than they should for the previously calculated mass of the galaxy, according to Newton's Law of Gravitation and Einstein's General Relativity Theory.

His conclusion was that there must be some sort of "dark matter" that did not emit light but added to the mass of the galaxies. Since then, numerous astronomers have performed similar experiments and have stated various theories on what is happening.

Using luminosity to measure mass

Luminosity is the amount of electromagnetic energy an astronomical body radiates per unit of time. It is independent of distance and related to the mass of the emitting body. Since the luminosity and mass of the Sun have been calculated, the mass-luminosity relation is handy for astronomers to use to determine the mass of a distant star or galaxy:

$$L/L_s = (M/M_s)^4$$

or

$$M = M_s[^4\sqrt{(L/L_s)}]$$

where

- $^4\sqrt{}$ is the 4th root (square root of the square root)
- M is the mass of the star or galaxy in kg
- M_s is the mass of the Sun = $1.988*10^{30}$ kg
- L is the luminosity of the star or galaxy in watts (W)
- L_s is the luminosity of the Sun = $3.84*10^{26}$ W

Substitute values for the Sun to get the formula:

$$M = (1.988*10^{30}/4.4*10^6)*[^4\sqrt{(L)}] \text{ kg}$$

$$M = (4.5*10^{23})*[^4\sqrt{(L)}] \text{ kg}$$

By measuring the luminosity of distant galaxies, Zwiky and other astronomers calculated their mass.

Using Doppler Effect to measure mass

The wavelength of a luminous object that is moving with respect to the observer changes according to the direction and velocity of the object. This phenomenon is known as the Doppler Effect.

When the object is moving away from the observer, the effect is known as a red-shift, since the wavelengths increase toward red. When moving toward the observer, it is called a blue-shift.

By using a spectrometer, the change in wavelength from the expected or normal spectrum can be measured and the velocity of the object calculated. Then the equation for the velocity of an orbiting object can be used to determine the center mass.

Velocity equation for planets in orbit

The velocity of planets orbiting the Sun can be approximated—assuming circular orbits—from the equation:

$$v_T = \sqrt{(GM_s/R)}$$

where

- v_T is the linear tangential velocity of the planet in orbit in km/s
- G is the Universal Gravitational Constant = $6.674*10^{-20}$ km³/kg-s²
- M_s is the mass at the Sun = $1.988*10^{30}$ kg
- R is the separation between the centers of the planet and Sun in km

(See chapter *12.4 Circular Planetary Orbits* for more information.)

This relationship shows that the further a planet is from the Sun the slower its velocity.

Applying to galaxies

The equation can also apply to galaxies, assuming that the mass of the galaxy is concentrated at its center of mass (CM).

> (See chapter *11.1 Overview of Gravitation and Center of Mass* for more information.)

Using the Doppler Effect to find the velocities of the various suns in the galaxy, the mass of the galaxy can be calculated by rearranging the above equation into:

$$M_g = v^2 R_g / G$$

where

- v is the velocity of a sun in the galaxy
- M_g is the mass of the galaxy at its center
- R_g is the separation between the CM and the center of the sun

Comparison of results

The suns at the outer edges of the galaxy should be moving slower than those closer toward the center. However, measurements of the velocities of the outer suns resulted in much higher values than expected for the mass determined from the luminosity measurements.

In order for the suns to remain in orbit for the measured velocities, the mass of the galaxy would have to be much higher. In fact, Zwiky's measurements came up with a mass 400 times of what was expected.

Other scientists confirmed the measurements. They estimate that 90 to 99 percent of the total mass of the universe is dark matter or matter that they cannot see.

Dark Energy

While dark matter seems to be adding to the gravitation in the Universe, dark energy seems to be a force pushing stars and galaxies apart.

The Universe appears to be expanding away from a center point, presumably the source of the Big Bang. Scientists and astronomers have calculated that the expansion would slow down with time, due to the mutual gravitational attraction between the galaxies.

However, this does not seem to be the case, as the Universe is actually expanding at a faster rate than before.

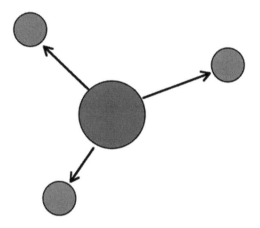

Dark energy seems to push objects apart

A possible explanation for the increased rate of expansion is a form of anti-gravitational force called dark energy. Unfortunately, dark energy has never been observed or measured. Instead it is an effort to explain the increased rate of expansion of the Universe.

Problems and alternatives

The idea of invisible matter and energy is troubling to scientists and astronomers. Some have established alternatives or flaws in the assumptions.

Alternatives to dark matter

One alternative to the existence of dark matter is that the added gravitational forces are actually caused by numerous Black Holes in the galaxies. Since light cannot escape Black Holes, they would be unseen. They also exhibit a large gravitational force.

A problem with the argument is that the Black Holes would also affect the orbits of any galaxies or suns that passed by, and that has not been observed.

(See chapter *13.5 Gravitational Escape Velocity for a Black Hole* for more information.)

Another alternative to dark matter is the *Modified Newtonian Dynamics Theory*, which proposes that at higher speeds or accelerations seen in stars at the outer edges of galaxies, gravitational attraction would fall off as a simple inverse of the separation instead of the inverse square of the separation in the Universal Gravitation Equation. This would allow stars on the outer edge of a galaxy to be held by a stronger gravitational pull.

One more concept is the existence of large quantities of particles, such as the neutrino, that do not readily interact with other forms of matter and are difficult to detect.

Problem with dark energy

Likewise, the existence of dark energy begs the question of why it only affects galaxies and not smaller objects of matter. Instead of being anti-gravitation, it could be a characteristic of space or perhaps some other unknown force that is applicable for only extremely large masses. Some feel that dark energy implies that the General Relativity Theory does not apply in certain situations.

Summary

Dark matter and dark energy affect gravitation in opposite ways.

Dark matter is invisible material that seems to add to the gravitation in galaxies. Dark energy seems to accelerate the expansion of the Universe with an anti-gravitation force. Both dark matter and dark energy are theories to explain anomalies in gravitation for objects at the galaxy scale of measurement.

Mini-quiz to check your understanding

1. How is the mass of a galaxy approximated from the velocity of its stars?

    a. From the equation $E = mv^2/2$

    b. By the equation for circular gravitational orbits

    c. Velocity affects luminosity and thus the mass

2. What is the reason for proposing dark energy?

    a. It is a way to explain the expansion of the Universe

    b. Something is necessary to counter dark matter

    c. It is a way to explain energy observed at night

3. What is an alternative theory to dark matter?

    a. Dark anti-matter

    b. Changes in Newton's laws for high speeds and large separations

    c. Matter comes in various colors and shades

Answers

1b, 2a, 3b

8.7 Gravitation as a Fundamental Force

Several theories state that there are four *Fundamental Forces* or *Interactions* that affect the way that objects or particles of matter interact with each other at a distance or separation. These forces are considered fundamental because they cannot be described in terms of other interactions.

Originally, the *Standard Model Theory* gave the fundamental forces as strong weak nuclear forces and electromagnetic force. However, the model did not address general relativity or gravitation.

The more recent *String Theory* and *Theory of Everything* added gravitation as a fundamental force.

The theories provide an explanation of how gravitation fits in the scheme of things and seem to resolve the disconnect between *General Relativity* and *Quantum Mechanics*. They explain forces as an exchange of fundamental particles.

Nuclear forces

Nuclear forces are divided into what they call Strong and Weak forces.

Strong force

The strong force is the attraction that holds the nucleus of an atom together, overcoming the repulsive electrical force of the positive (+) charged protons. The relative strength of the strong force is designated as 1.

The range of this force is small, approximately the diameter of a medium-sized nucleus (10^{-15} m). Apparently, this force does not decrease by the inverse square as do the gravitational and electromagnetic forces. Instead, it just stops at its given separation.

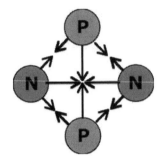

Strong force holds nucleus together

The *Theory of Exchange Forces* designates the *gluon* as the "glue" that holds the nucleus together, through some sort of exchange between nuclear particles.

Weak force

The weak force in a nucleus involves an exchange of *W and Z vector boson* particles. The strength of the weak force is 10^{-6} that of the strong force. Its range is only 10^{-18} m, which is about 0.1% of the diameter of a proton.

The apparent purpose of the weak nuclear force is to allow deuterium fusion to take place. This is necessary for our Sun and the stars to burn. Deuterium is a hydrogen isotope.

The force is also necessary for the creation of heavy nuclei and causes phenomena such as beta decay.

Electromagnetic force

Electromagnetic force consists of the attraction and repulsion of materials consisting of electric charges, as well as magnetic materials.

Strength and range

The relative strength of the electromagnetic force is 1/137 of the strong nuclear force. Its force drops off as the square of the separation between charged particles or magnetic poles, although the range is infinite.

Electric force

Electric charges can be positive (+) or negative (−), where like charges repel and unlike charges attract. Protons have a positive (+) charge and electrons have a negative (−) charge. Electric forces are what hold atoms and molecules together.

Magnetic force

Magnetic poles can be north (or north-seeking) and south (or south-seeking). Like poles repel and unlike poles attract. Moving and spinning electrical charges create a magnetic field, depending on their direction of motion.

Cause

At one time electromagnetic forces were explained as a property of space that consisted material called *aether*. The present theories explanation is that the force is caused by the exchange of *photon* particles.

Gravitational force

Gravitational force is the attraction between objects. It can be compared with the other three interactions or forces.

Range and strength

Gravitation has at reach or range to infinity. However, it is the weakest of the fundamental forces. The gravitational strength is only $6*10^{-39}$ of the strength of the strongest nuclear force.

> **Note**: 10^{-39} equals $1/10^{39}$, where 10^{39} is 1 followed by 39 zeros. That is a very small number.

The strength of the gravitational force decreases as the square of the separation between two objects, as does the electromagnetic force. The nuclear forces do not have that feature.

Although the gravitational force is much smaller than the other fundamental forces, it's impact concerns objects of large mass, such as planets and stars. Gravitation is what keeps the Earth and other planets in orbit around the Sun, as well as the Universe in order.

Explanations of gravitation

The *Theory of General Relativity* explains the force due to gravitation as a result of the curvature of space caused by matter.

> (See chapter *8.4 General Relativity Theory of Gravitation* for more information.)

The *Theory of Quantum Mechanics* explains gravitation as caused by the exchange of *graviton* particles between the masses. Graviton particles travel at the speed of light. Also, the wave-particle duality of Quantum Mechanics means that gravitation waves are possible.

> (See chapter *8.5 Quantum Theory of Gravitation* for more information.)

Effect of dark energy

Although gravitation is only an attractive force, some scientists speculate that there may be a sort of anti-gravitation that causes objects to repel away from each other.

This has been measured in the rate of expansion of the Universe, which is increasing, as opposed to decreasing due to gravitation. The force causing the increase in expansion is called *dark energy*.

> (See chapter *8.6 Effect of Dark Matter and Dark Energy on Gravitation* for more information.).

Summary

The four Fundamental Forces that affect the way that objects or particles of matter interact with each other at a separation are strong nuclear, weak nuclear, electromagnetic and gravitational forces. They are part of the Theory of Everything that is an effort to resolve the differences in relativity and quantum mechanics.

Gravitation is the weakest of the fundamental forces, has an infinite range and is apparently implemented by an exchange of graviton particles.

Mini-quiz to check your understanding

1. How are the strengths of the fundamental forces compared?

    a. The strong nuclear force is designated as 1 and the others are fractions of that force

    b. Each is measured as they compare to gravitation

    c. All fundamental forces have the same strength

2. Which fundamental force has the same range as gravitation?

    a. Strong nuclear force

    b. All of the other forces

    c. Electromagnetism

3. How fast does gravitation travel?

    a. It has infinity speed

    b. The speed of light

    c. 9.8 m/s^2

Answers

1a, 2c, 3c

Part 9: Gravitation Principles

Applications of the various theories and equations of gravitation are explained in this part.

Part 9 chapters

Chapters in Part 9 include:

9.1 Equivalence Principles of Gravitation

The Weak Equivalence Principle, equivalence of inertial and gravitational mass and the Strong Equivalence Principle are explained in this chapter.

9.2 Similarity Between Gravitation and Electrostatic Forces

This chapter shows how are the gravitational force and electrostatic force equations similar and are different. It also delves into gravitomagnetism.

9.3 Gravitational Speed

This chapter explains Newton's view on the speed of gravitation and compares it with speed from relativity and according to quantum physics concepts?

9.4 Gravitational Potential Energy

This chapter explains what gravitational potential energy is and gives the derivation of the potential energy equation. It also explains about kinetic and total energies.

9.1 Equivalence Principles of Gravitation

There are several *Equivalence Principles* that refer to related gravitational concepts.

The *Weak Equivalence Principle* states that objects fall at the same rate, provided that are freely falling. The *equivalence of inertial and gravitational mass* states that mass determined by inertia is the same as mass determined by gravitation.

The *Strong Equivalence Principle* extends the equivalence of masses to state that observations of acceleration cannot be distinguished from gravity.

Weak Equivalence Principle

The *Weak Equivalence Principle* (also called the *Uniqueness of Free Fall Principle*) states that gravitational causes objects to fall or move toward an attracting body at the same rate, independent of their mass.

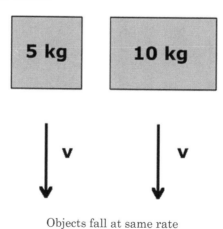

Objects fall at same rate

Proof

The proof of this principle is pretty straightforward. Consider two objects that are the same separation from a larger body. Their equations are:

$$F_1 = Gm_1 M/R^2$$

$$F_2 = Gm_2 M/R^2$$

where

- F_1 and F_2 are the forces on objects 1 and 2 respectively
- G is the Universal Gravitational Constant
- m_1 and m_2 are the masses of objects 1 and 2 respectively
- M is the mass of the attracting body
- R is the separation from the centers of the objects to the center of the attracting body

Since $F = ma$, the acceleration is GM/R^2 and is the same for both objects. Thus, they will fall at the same rate.

Restrictions

However, there are some restrictions on this principle.

No outside forces

It is assumed that there are no outside forces such as air resistance acting on the falling objects. In other words, they are falling freely.

Mass much less than attracting body

A major restriction on the Weak Equivalence Principle is that the mass of each falling object must be much less than that of the attracting body.

The gravitational force causes both the falling object and the attracting body to move toward each other and their center of mass. Thus, the mass of the falling object much be so small with respect to the attracting body that its movement is negligible.

(See chapter *11.1 Overview of Gravitation and Center of Mass* for more information.)

For example, the mass of the Earth is $5.974*10^{24}$ kg. An object that had a mass of 6,000,000 kg ($6*10^6$ kg) falling from a displacement of 10^5 km would result in movement of the Earth of:

$$r_M = mR/(M + m) \text{ km}$$

where

- r_M is the separation between the center of the attracting body and the center of mass between the objects
- m is the mass of the smaller object
- R is the separation between the objects
- M is the mass of the larger, attracting object

Thus:

$$r_M = (6*10^6)10^5/(5.974*10^{24} + 6*10^6) \text{ km}$$

$$r_M = 5.974*10^{11}/6*10^{24} \text{ km}$$

This is approximately:

$$r_M = 10^{-13} \text{ km} = 10^{-5} \text{ cm}$$

That is a tiny movement for a mass of that size.

Objects must be of similar size

Another restriction is that the objects must be similar in physical size, such that the center of mass for each is at approximately the same displacement from the attracting body.

If the separations between the centers of mass are different, the objects would fall at slightly different rates

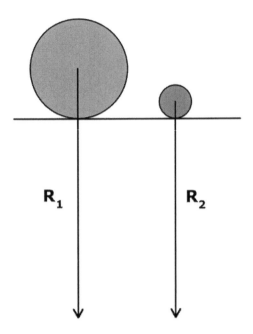

Exception when objects are much different in size

This exception is seldom considered when studying the principle.

Equivalence of inertial and gravitational mass

There is an equivalence of inertial and gravitational mass. You can see this by examining the forces from both inertial mass and gravitational mass.

Inertial mass

If you accelerate an object, the force required to overcome its inertia is:

$$F_i = m_i a$$

and the inertial mass is:

$$m_i = F_i/a$$

where

- F_i is the force needed to overcome inertia
- m_i is the inertial mass
- a is the acceleration on the object

Gravitational mass

Likewise, the gravitational force is:

$$F_g = Gm_gM/R^2$$

and the gravitational mass is:

$$m_g = F_gR^2/GM$$

where

- F_g is the gravitational force on the object
- G is the Universal Gravitational Constant
- m_g is the gravitational mass of the object
- M is the mass of the attracting object
- R is the separation between the objects, as measured from their centers of mass

Equivalence

Since the time of Newton, scientists have wondered if the inertial mass was the same as the gravitational mass. Does $m_i = m_g$? Many experiments verify the equivalence.

Albert Einstein stated that a gravitational force, as experienced locally while on a massive body such as the Earth, is actually the same as the pseudo-force experienced by an observer in an accelerated frame of reference.

Einstein used the equivalence of inertial and gravitational mass as a basic framework for the *General Theory of Relativity*.

Strong Equivalence Principle

The *Strong Equivalence Principle* (also known as the *Einstein Equivalence Principle*) states that the effects of acceleration are indistinguishable from those of gravitation.

(See chapter *6.8 Artificial Gravity* for an example of this.)

Experiments by observer

This means that an observer cannot determine by experiment whether he or she is accelerating or in a gravitational field. In other words, results from experiments in an accelerating spaceship would be the same as those obtained from gravitation.

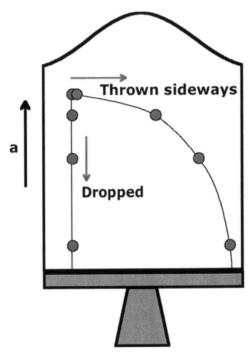

Experiment in accelerating spaceship

Note: One problem with this concept is that acceleration cannot be applied for too long a period, because the spaceship would soon reach the speed of light. On the other hand, gravitation is continuously present.

Einstein's conclusion

Einstein concluded that gravitation and motion through space-time are related and that the Strong Equivalence Principle suggests that gravitation is geometrical by nature.

Difference between strong and weak

The difference between the Strong Principle of Equivalence and the Weak Principle of Equivalence is that strong equivalence states all the laws of nature are the same in a uniform static gravitational field and the equivalent accelerated reference frame, while weak equivalence states all the laws of motion for freely falling particles are the same as in an unaccelerated reference frame.

Summary

The *Weak Equivalence Principle* states that objects fall at the same rate, provided that are much smaller than the attracting body and are freely falling.

The *equivalence of inertial and gravitational mass* states that mass determined by inertia is the same as mass determined by gravitation.

The *Strong Equivalence Principle* extends the equivalence of masses to state that observations of acceleration cannot be distinguished from gravitation.

Mini-quiz to check your understanding

1. What is a requirement for the Weak Equivalence Principle?

    a. Falling objects must be close to the mass of the attracting object

    b. The objects fall freely

    c. The force of gravity must be very weak

2. If you double the acceleration of an object, how would that affect its inertia mass?

    a. The acceleration has no effect on the mass of the object

    b. The mass would double, since it is proportionate to the acceleration

    c. The mass would become gravitational mass, provided **a = g**

3. What an an application of the Strong Equivalence Principle?

    a. There are no applications, since it is just a theory

    b. Time can be made to stand still

    c. You can simulate gravity in a spaceship

Answers

1b, 2a, 3c

9.2 Similarity Between Gravitation and Electrostatic Forces

Since the 1700s, scientists noticed the similarity between gravitation and electrostatic forces. This led to speculation that the two forces were somewhat related. However, there are also differences that separate the forces.

In the early 1900s, equations indicated that moving charges and moving masses provide an analogy of gravitational and electromagnetic fields. This results in the concept of a *gravitomagnetic force*, where large rotating masses affect nearby small masses. However, measurements have not validated the concept.

Force equations parallel

The gravitational force equation and the electrostatic force equation can be seen as parallel to each other.

Gravitation

The *Universal Gravitation Equation* states the force of attraction between two objects, where the mass is considered concentrated at their centers of mass:

$$F = GMm/R^2$$

where

- **F** is the force of attraction between two objects
- **G** is the Universal Gravitational Constant
- **M** and m are the masses of the two point objects
- **R** is the separation between the centers of the objects

Electrostatic

The electrostatic force equation is called *Coulomb's Law* and states the force of attraction between particles of opposite electrical charge. It also represents the force of repulsion for like charges:

$$F = k_e qQ/r^2$$

where

- **F** is the force of attraction or repulsion between two electrically charged particles
- k_e is the Coulomb force constant
- **q** and **Q** are point charges of the two particles
- **r** is the separation between the particles

Comparison of forces

There are similarities and differences between the two forces.

Similarities

From looking at the two force equations, you can see the similarities and how gravitational force can be considered parallel to the force between two charges.

Besides being proportional to the inverse of the square of the separation, both forces extend to infinity. They also both travel at the speed of light.

Differences

One major difference is in the strength of the forces. The gravitational attraction between two electrons is only $8.22*10^{-37}$ of the electrostatic force of repulsion at the same separation. However, gravitation usually is concerned with large masses, while any large collection of charges will quickly neutralize.

Another difference between the two forces is the fact that gravitation only attracts, while electrical forces attract when the electrical charges are opposite and repel if the charges are similar.

Thus, gravitation is considered a monopole force, while electrostatics is a dipole force.

However, the concept of *dark energy*, which seems to have an anti-gravitation force, may allow for gravitation to be both the attraction and repulsion of matter.

Gravitomagnetism

An analogy of gravitational and electromagnetic fields is seen by comparing the *Einstein Field Equations* from the *General Theory of Relativity* with *Maxwell's Field Equations* for electrical and magnetic fields.

Einstein's equations state that the gravitational field produced by a rotating object can be described by equations that have the same form as Maxwell's equations.

> **Note**: Both sets of equations are beyond the scope of our material.

Just as a rotating or moving electrical charge creates a magnetic field, general relativity predicts that a huge rotating mass will cause a small nearby free-falling object to rotate. This is called a *gravitomagnetic effect*. Unfortunately, this prediction of general relativity has yet to be directly tested or proven to be true

Summary

There is a similarity between gravitation and electrostatic force equations. Although this led to speculation that the two forces were somewhat related, differences were seen that separated the forces.

Einstein's and Maxwell's equations indicate that moving charges and moving masses provide an analogy of gravitational and electromagnetic fields, called gravitomagnetics. The result is that large rotating masses affect nearby small masses. However, measurements have not validated the concept.

Mini-quiz to check your understanding

1. How are the gravitation and electrostatic for equations parallel?

   a. They are in the same form but with different variables

   b. Either one could be used when making calculations

   c. Both are dependent on the mass of the objects

2. What might allow anti-gravitation?

   a. Increasing the mass of each object

   b. Reduction in electrical charge

   c. Dark energy

3. What does gravitomagnetism predict would happen to a free-falling marble near the Sun?

   a. The marble would burn up, due to the Sun's heat

   b. The marble should start to spin due to the Sun's rotation

   c. The marble would be repelling into space, due to the Sun's anti-gravity

Answers

1a, 2c, 3b

9.3 Gravitational Speed

Change in the separation of two objects results in a change in the gravitational force between them. *Gravitational speed* concerns the amount of time it takes the gravitational force to accommodate the change in separation over the given displacement. It is often carelessly called the *speed of gravity*.

Concepts about the mechanism of gravitation determine what scientists say is its reaction time or speed. The Newtonian model for gravitation states that the change in force is instantaneous and that the gravitational speed is infinite. However, some astronomical observations required a better explanation.

The *Theory of General Relativity* stated that gravitation travels at the speed of light and is caused by the curvature of spacetime. The *Quantum Theory of Gravitation* states that gravitation is caused by the exchange of *graviton* particles, which travel at the speed of light.

Speed from Law of Universal Gravitation

In establishing the *Law of Universal Gravitation*, Isaac Newton said in 1687 that he was uncertain about the mechanism of gravitation, except that it was some sort of action-at-a-distance.

His *Universal Gravitation Equation* shows the gravitational force between two objects at a given instance:

$$F = GMm/R^2$$

where

- **F** is the force of attraction between two objects
- **G** is the Universal Gravitational Constant
- **M** and **m** are the masses of the two objects
- **R** is the separation between the objects, as measured from their centers of mass

---

Movement of one object, with respect to the other, results in a change in separation and thus a change in the force between the objects.

Infinite speed

In order to satisfy *Kepler's Laws of Planetary Motion*—especially for elliptical orbits—Newton came to the conclusion that the gravitational force propagates instantaneously, irrespective of the separation between objects.

In other words, Newton's concepts state that the gravitational speed is infinite.

For example, the changing displacements due to elliptical orbits for any planets in the Solar System would make the orbits unstable if the gravitational force did not propagate instantaneously.

However, instantaneous propagation or infinite speed is counterintuitive to the way things work in the Universe. This had bothered scientists over the years. But the assumption was adequate to account for astronomical phenomena with the observational accuracy of those days.

Searching for better solution

In 1847, French astronomer Urbain Le Verier determined that the elliptical orbit of Mercury—the planet closest to the Sun—has a precession that was at a significantly different rate than was predicted by Newton's theory.

In searching for a solution scientists tried a new mechanism for gravitation, combining Newton's force law with the established laws of electrodynamics, placing the gravitational speed equal to the speed of light.

Unfortunately, those theories provided insufficient.

Speed from relativity

In 1915, Albert Einstein's *Theory of General Relativity* took a completely different approach, stating the gravitation was not a force but a result of the curvature of spacetime.

With this mechanism of gravitation, a change in separation resulted in a change of the curvature of spacetime between the objects.

Since the *Special Theory of Relativity* stated that the speed of light, **c**, was a fundamental constant and was the ultimate speed for any physical interaction, the conclusion was that the gravitational speed was the speed of light.

In other words, the change in curvature as a result of the change in separation of the object occurred at the speed of light. This also presented the possibility of gravitation waves.

Equations in the relativity theories showed that Kepler's Laws of Planetary Motion were approximations and explained the elliptical orbit of Mercury.

Speed from quantum gravitation

In 1927, *Quantum Physics* formalized the various areas of *Quantum Mechanics* into one field. However, there was a disconnect between Quantum Physics and General Relativity.

The recent development of *Quantum Gravitation*, along with *String Theory* and the *Theory of Everything* are efforts to unify Quantum Physics and Relativity.

The Quantum Gravitation mechanism for the gravitational force is the exchange of *graviton* particles, which travel at the speed of light. From the Quantum Physics wave-particle duality concept, those graviton particles could also be represented as gravitational waves.

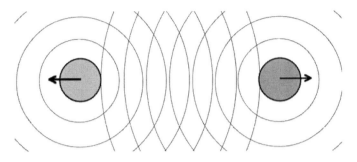

Gravitation waves responding to change in separation

Summary

Gravitational speed concerns the amount of time it takes the gravitational force to accommodate the change in separation over the given displacement. The mechanism of gravitation is a determining factor in its reaction time or speed.

The Law of Universal Gravitation did not define a mechanism for gravitation but stated states that the change in force is instantaneous and that the gravitational speed is infinite.

The Theory of General Relativity stated that gravitation travels at the speed of light and is caused by the curvature of spacetime. The Quantum Theory of Gravitation states that gravitation is caused by the exchange of graviton particles, which travel at the speed of light.

Mini-quiz to check your understanding

1. Why did Newton decide that the effect of gravitation was instantaneous?

    a. To satisfy Einstein's theories

    b. To satisfy Mercury's orbit

    c. To satisfy Kepler's law

2. How does the curvature of space respond to a change in separation between planets?

    a. The curvature changes according to separation of masses

    b. The curvature causes to planets to stay in one place

    c. The curvature is fixed and does not change

3. How fast do graviton particles move?

    a. They move at the speed of falling bodies

    b. They don't move but remain fixed within a mass

    c. They move at the speed of light

Answers

1c, 2a, 3c

9.4 Gravitational Potential Energy

The *gravitational potential energy* between two objects of mass is the potential of motion caused by their gravitational attraction. The attraction of the objects turns the potential energy into the kinetic energy of motion, such that the objects will move toward each other.

Potential energy of two objects at a given separation is defined as the work required to move the objects from a *zero reference point* to that given separation. The force required to move the objects equals the incremental change of the potential energy for a change in separation.

Combining the incremental change equation with the *Universal Gravitation Equation* and then integrating from infinity to the given separation results in the equation for the potential energy. The kinetic energy and total energy can lead to the gravitational escape velocity equation and other applications.

Work, potential energy and force

At a given separation, the gravitational potential energy (**PE**) between two objects is defined as the work required to move those objects from a zero reference point to that given separation.

Work

Work is often defined as the product of the force to overcome a resistance and the displacement of the objects being moved.

When you move two objects apart, you do work to overcome the gravitational resistance. However, when the objects are allowed to move toward each other, the work is "given back" and is considered negative work.

Zero reference point

The zero reference point is where the potential energy is considered to be null. In the case of gravitation between two objects, that zero reference point is at a hypothetical infinite separation.

> **Note**: In the case of the potential energy from gravity, the surface of the Earth is the zero reference point, and the potential energy is the work required to lift the object to some height **PE = mgh**.
>
> (See chapter *6.1 Potential Energy of Gravity* for more information.)

The work done to move two objects apart against the gravitational force to an infinity separation is positive work, while the work in the direction of gravitation is negative work.

Force used instead of work

Since the zero reference point for gravitation is at infinity, the displacement is infinite. This makes the calculation of potential energy as a function of work difficult.

Thus, it is more convenient to use the fact that the gravitational force between two objects is equivalent to the incremental change in potential energy with respect to a change in separation:

F = dU/dR

where

- **F** is the gravitational force in newtons (N)
- **dU** is the derivation or incremental change in potential energy in joules (J)
- **dR** is the incremental change in separation in meters
- **dU/dR** is the first derivation of **U** with respect to **R**

Note: Both **PE** and **U** are commonly used to denote potential energy. Right now, we are using **U**, since it is more convenient to state **dU** as the derivative of the potential energy.

This equation is used to find the gravitational potential energy.

Derivation of PE equation

The derivation of the gravitational potential energy equation starts with the *Universal Gravitation Equation*:

$$F = GMm/R^2$$

where

- **G** is the Universal Gravitational Constant
- **M** and **m** are the masses of the objects
- **R** is the separation between the centers of mass of the objects

Note: Each object has a center of mass. However, there is also a center of mass of the system, which is considered when dealing with kinetic energy.

Combining this equation with the previous increment equation, you get:

$$dU/dR = GMm/R^2$$

$$dU = (GMm/R^2)dR$$

Since at $R = \infty$, the potential energy is zero, you integrate over the range of **U = U** to **U = 0** and from **R = R** to $R = \infty$:

$$\int dU = \int (GMm/R^2)dR$$

The result is:

$$U = -GMm/R$$

or

$$PE = -GMm/R$$

Since the gravitational potential energy between two objects is defined as the work required to move those objects from a zero reference point to that given separation, and since that work is negative, the potential energy is also negative.

Energy and applications

The force of gravitation attracts objects toward each other, so both will also have kinetic energy at any separation unless there is some external force to hold them apart.

Assuming that the only force is that of gravitation between the objects, the system is considered closed and the total energy remains constant, according to the *Law of Conservation of Energy*.

This can lead to the escape velocity equation and has implications in orbiting objects.

Kinetic energy

The kinetic energy of the two objects is stated in the equations:

$$KE_M = MV^2/2$$

$$KE_m = mv^2/2$$

where

- KE_M and KE_m are the kinetic energies of the objects in joules (J)
- **M** and **m** are the masses of the objects in kg
- **V** and **v** are the velocities of the objects toward the center of mass between them in m/s

Note: The velocities of the objects are relative to the center of mass between them. The relationship of those velocities follows if **M > m** (**M** is larger than **m**), then **V < v**.

(See chapter *11.1 Gravitation and Center of Mass* for more information.)

Total energy

According to the *Law of Conservation of Energy*, the total energy (**TE**) of a closed system remains constant:

$$TE = (KE_M + KE_m) + PE$$

Values of **R** will determine values of **V**, **v** and **PE** accordingly.

Escape velocity

In the case where the mass of one object, **M**, is much greater than the mass of the other object (**M >> m**), the velocity of the larger object can be considered negligible and its kinetic energy equal to zero:

$$KE_M = 0$$

In such a case, the total energy equation is approximately:

$$TE = KE_m + PE$$

$$TE = mv^2/2 - GMm/R$$

Since the total energy at infinity is zero:

$$mv^2/2 = GMm/R$$

This leads to the escape velocity equation:

$$v_e = -\sqrt{(2GM/R_i)}$$

where

- v_e is the escape velocity vector
- R_i is the initial separation between the two objects

Note: v_e is negative, since it is moving away from the center of mass between the objects

(See chapter *13.2 Gravitational Escape Velocity Derivation* for more information.)

Orbits

In the case where the objects are rotating about the center of mass between them, they have a potential energy according to their separation, but their kinetic energy in the direction of the line between their centers is zero.

This means that although the objects are moving in orbit, there is no work against gravitation.

> (See chapter *12.4 Circular Planetary Orbits* for more information.)

Summary

The gravitational force of attraction between objects at some separation creates their gravitational potential energy, which can be turned into the kinetic energy of motion. Potential energy is defined as the work required to move the objects from a zero reference point to a separation in space.

Gravitational force equals the incremental change of the potential energy for a change in separation and is used to derive the potential energy equation. Resulting kinetic energy and total energy can lead to applications such as gravitational escape velocity equation.

Mini-quiz to check your understanding

1. Where is the zero reference point for gravitation?

    a. At the surface of the Earth

    b. At infinity

    c. At zero

2. Why is the gravitational potential energy a negative number?

    a. It is decreasing as you get further from the zero reference point

    b. It is always less than the kinetic energy

    c. It is the escape velocity

3. What are two applications of the combination of **PE** and **KE**?

    a. There are no applications, since they are different forms of energy

    b. Force and work

    c. Escape velocity and orbits

Answers

1b, 2a, 3c

Part 10: Gravitation Applications

Applications of the various theories and equations of gravitation are explained in this part.

Part 10 chapters

Chapters in Part 10 include:

10.1 Gravitational Force Between Two Objects

The Universal Gravitation Equation is used to calculate the force between the Earth and the Moon, a girl and the Moon and a girl and a boy.

10.2 Cavendish Experiment

The Cavendish experiment is a way to determine the Universal Gravitational Constant **G**. This chapter gives the background on the experiment, shows the experimental method and provides the derivation of the value of **G**.

10.3 Influence of Gravitation in the Universe

This chapter shows how gravitational forces played in forming galaxies and stars, why the stars and planets are round and what keeps the planets in orbit.

10.4 Gravitation Causes Tides on Earth

This chapter shows how tides are caused by the gravitational forces from the Moon and Sun that attract the ocean water toward them and away from other areas in the ocean.

10.1 Gravitational Force Between Two Objects

You can apply the *Universal Gravitation Equation* to show the force of attraction between two objects, provided you know the mass of each object and their separation. The equation is:

$$\mathbf{F = GMm/R^2}, \text{ where } \mathbf{G} = 6.674*10^{-11} \text{ N-m}^2/\text{kg}^2.$$

With this equation, you can make calculations to determine such things as the force between the Earth and the Moon, between a boy and a girl and between the Moon and a girl.

Force attracting Earth and Moon

To calculate the gravitational force pulling the Earth and Moon together, you need to know their separation and the mass of each object.

Separation

The Earth and Moon are approximately $3.844*10^5$ kilometers apart, center to center. Since the units of $\mathbf{G}$ are in meters, you need to change the units of separation to meters.

$$\mathbf{R} = 3.844*10^8 \text{ m}$$

Mass of each object

Let $\mathbf{M}$ be the mass of the Earth and $\mathbf{m}$ the mass of the Moon.

$$\mathbf{M} = 5.974*10^{24} \text{ kg}$$

$$\mathbf{m} = 7.349*10^{22} \text{ kg}$$

Force of attraction

Thus, the force of attraction between the Earth and Moon is:

$$F = GMm/R^2$$

$$F = (6.674*10^{-11} \text{ N-m}^2/\text{kg}^2)(5.974*10^{24} \text{ kg})$$
$$(7.349*10^{22} \text{ kg})/(3.844*10^8 \text{ m})^2$$

$$F = (2.930*10^{37} \text{ N-m}^2)/(1.478*10^{17} \text{ m}^2)$$

$$F = 1.982*10^{20} \text{ N}$$

Note: Notice how all the units, except N, cancelled out.

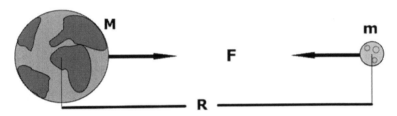

Attraction between Earth and Moon

Result of force

This considerable force is what holds the Moon in orbit around the Earth and prevents it from flying off into space. Inward force of gravitation equals the outward centrifugal force from the motion of the Moon.

> (See chapter *12.4 Circular Planetary Orbits* for more information.)

Also, the gravitational force from the Moon pulls the oceans toward it, causing the rising and falling tides, according to the Moon's position.

> (See chapter *10.4 Gravitation Causes Tides on Earth* for more information.)

Force attracting boy and girl

If a boy who weighed 75 Newtons (kg-force) or 165 lb sat 0.5 m (19.7 in) from a 50 kg (110 lb) girl, what would be the gravitational attraction between them?

Note: It is common—but scientifically incorrect—to state a person's weight in kg, which is mass. Care must be taken to distinguish between units of weight and mass. Newtons or kg-force are weight and kg or kg-mass are units of mass.

Also note: The separation between the boy and girl is measured from the center of the first person to the center of the other.

In order to establish the force between the two people, their weights must be converted to units of mass. That means dividing each weight by 9.8 m/s^2. Thus, the mass of the boy is:

M = 75/9.8 = 7.653 kg-mass

The mass of the girl is:

m = 50/9.8 = 5.102 kg-mass

Substituting the values into the equation, you get:

F = GMm/R^2

F = $(6.674*10^{-11})(7.653)(5.102)/(0.5)^2$ N

F = $260.59*10^{-11}/0.25$ N

F = $1040.361*10^{-11}$ N

This is approximately:

F = 10^{-8} N or 0.01 millionth of a newton

That is a very small gravitational attraction, but it can be measured on a sensitive instrument.

Force attracting girl and Moon

What is the gravitational pull from the Moon on the 50 kg (110 pound) girl?

You need to find the separation between the Moon and the surface of the Earth. Since the separation between the centers of the Earth and Moon is $3.84*10^8$ m, you subtract the radius of the Earth (6371 km) to get the distance to the Earth's surface:

6371 km = 6,371,000 m = $0.064*10^8$ m

$\mathbf{R} = 3.844*10^8$ m $- 0.064*10^8$ m $= 3.780*10^8$ m

Thus, the force between the girl and the Moon is:

$\mathbf{F = GMm/R^2}$

$\mathbf{F} = (6.674*10^{-11})*(5.102)*(7.349*10^{22})/(3.780*10^8)^2$ N

$\mathbf{F} = 250.239^{11}/14.228*10^{16}$ N

$\mathbf{F} = 17.588*10^{-5}$ N

$\mathbf{F} = 1.759*10^{-4}$ N $= 0.000176$ N

The girl would not notice the pull from the Moon, since the gravitation pull on her toward the Earth is 50 N (50 kg-m/s^2 or kg-weight), which is much larger.

But still, she is attracted more toward the moon than toward the boy who was sitting next to her.

Summary

You can apply the Universal Gravitation Equation to show the force of attraction between two objects. With this equation, you can show the force between the Earth and the Moon is $\mathbf{F} = 1.982*10^{20}$ N, between a boy and a girl is $\mathbf{F} = 10^{-8}$ N and between the Moon and a girl is $\mathbf{F} = 1.759*10^{-4}$ N.

Mini-quiz to check your understanding

1. What is one result of the great force between the Earth and the Moon?

   a. Oceans are pulled toward the Moon

   b. There is no gravitation on the Moon

   c. Items often will just fly into space

2. What happens to the gravitational attraction between a boy and a girl as they move closer together?

   a. The force of attraction goes up, because the separation is smaller

   b. They are told not to sit so close to each other

   c. The force of attraction goes down, because the separation is smaller

3. Why is the girl attracted more to the Moon than the boy next to her?

   a. She likes the Moon and dislikes the boy

   b. The distance is greater to the Moon and thus the force is greater

   c. The Moon has a much greater mass than the boy

Answers

1a, 2a, 3c

10.2 Cavendish Experiment

A major element in the Universal Gravitation Equation, $F = GMm/R^2$, is the Universal Gravitational Constant, **G**. The constant was not determined until many years after Isaac Newton formulated his equation, as a result of what is called the Cavendish experiment.

This experiment used a torsion balance device to attract lead balls together, measuring the torque on a wire and equating it to the gravitational force between the balls. Then by a complex derivation, the value of **G** was determined.

Background of experiment

After Isaac Newton formulated the Universal Gravitation Equation in 1687, there really wasn't much interest in **G**. Most scientists simply considered it a proportionality constant. They were more interested in gravity than in gravitation.

Cavendish measures density of Earth

In 1798, Henry Cavendish performed an experiment to determine the density of the Earth, which would be useful in astronomical measurements. He used a torsion balance invented by geologist John Mitchell to accurately measure the force of attraction between two masses.

From this measurement, he determined the mass of the Earth and then its density. In Cavendish's published paper on the experiment, he gave the value for the density and mass of the Earth but never mentioned the value for **G**.

Others determine G

It wasn't until 1873 that other scientists repeated the experiment and documented the value for **G**. The value for **G** implied from Cavendish's experiment was very accurate and within 1% of present-day measurements.

Cavendish given credit

Because his experiment ultimately determined the value for **G**, Cavendish has been often incorrectly given credit for determining the gravitational constant.

Cavendish experiment setup

The Cavendish experiment uses a torsion balance to measure the weak gravitational force between lead balls. A torsion balance consists of a bar suspended from its middle by a thin wire or fiber. Twisting the fiber requires a torque that is a function of the fiber width and material.

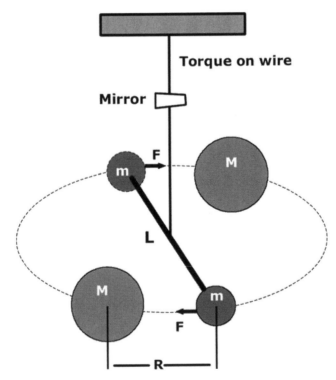

Cavendish experiment to measure gravitation

Gravity and Gravitation

When the gravitational force pulls the balls on the bar toward the stationary balls, the bar turns until the torque of the fiber balances the gravitational force. The magnitude of the force can then be calculated from the angle through which the bar turned. This angle is accurately determined by a mirror placed on the fiber.

However, the inertia of the balls causes them to go slightly beyond the balance point and thus create a harmonic oscillation around that point. This can also be measured by the light reflected from the mirror. The rate of oscillation is used to determine the spring constant of the wire, which is necessary in the final calculation of G.

It is truly a clever experiment

The most recent measurement used a device called an atomic interferometer to measure **G**.

Derivation of G

The derivation of the value for **G** from the experiment is fairly complex. Variables measured in the experiment are:

M is the mass of the larger ball in kg

m is the mass of the smaller ball in kg

R is initial separation between the balls in meters

L is the length of the balance bar in meters

θ (small Greek letter omega) is the angle from the rest position to the equilibrium point measured in radians

T is the oscillation period in seconds

The Universal Gravitation Equation for the balls is:

$$F = GMm/R^2$$

where

- **F** is the force of attraction between the balls in newtons (N)
- **G** is the Universal Gravitational Constant in in N-m^2/kg^2 or m^3/kg-s^2

Solving for **G**:

G = FR^2/Mm

In order to get a value for **G**, the force must be determined.

Force related to torque

The force **F** is related to the torque on the fiber. The equation for torque is the applied force times the moment arm. Since there are two moment arms of **L/2**, the torque is:

$\tau = FL$

where τ (small Greek letter tau) is the torque in N-m. Thus:

F = τ/L

Torque related to torsion coefficient

However, the torque is also related to the torsion coefficient of the fiber or wire:

$\tau = \kappa\theta$

where κ (small Greek letter kappa) is the torsion coefficient in newton-meters/radian. Thus:

F = $\kappa\theta$/L

Torsion coefficient related to oscillation period

The unknown factor is the torsion coefficient, which is calculated by measuring the resonant oscillation period of the wire.

When the balance bar is initially released and the moving balls approach the larger balls, the inertia of the smaller causes them to overshoot the equilibrium angle. This results in the torsion balance oscillating back-and-forth at its natural resonant oscillation period:

$$T = 2\pi\sqrt{(I/\kappa)}$$

where

- **T** is the oscillation period in seconds
- π (small Greek letter pi) is 3.14...
- **I** is the moment of inertia of the smaller balls in kg-m^2

Note: the mass of the bar is considered negligible and not a factor in the inertia.

Oscillation period related to moment of inertia

The moment of inertia of the smaller balls is:

$$I = mL^2/2$$

Substitute inertia in the torque equation:

$$T = 2\pi\sqrt{(mL^2/2\kappa)}$$

Solve for κ:

$$T^2 = 4\pi^2(mL^2/2\kappa)$$

$$2\kappa T^2 = 4\pi^2 mL^2$$

$$\kappa = 2\pi^2 mL^2/T^2$$

Substitute κ in the equation for **F**:

$$F = \kappa\theta/L$$

Thus:

$$F = 2\pi^2 mL^2\theta/LT^2$$

Find G

Substitute for **F** in the equation for **G**:

$$G = FR^2/Mm$$

$$G = 2\pi^2 mL^2\theta R^2/LT^2Mm$$

Simplify the equation:

$$G = 2\pi^2 L\theta R^2/T^2M$$

The calculated value of **G** from this experiment is:

$$G = 6.674*10^{-11} \text{ m}^3/\text{kg-s}^2$$

Since a newton is equivalent to kg-m/s^2, **G** also is defined as:

$$G = 6.674*10^{-11} \text{ N-m}^2/\text{kg}^2$$

Summary

Henry Cavendish performed an experiment to find the density of the Earth. Other scientists used his experimental setup to determine the value of **G**.

The setup consisted of a torsion balance to attract lead balls together, measuring the torque on a wire and then equating it to the gravitational force between the balls. Then by a complex derivation, the value of **G** was determined.

Mini-quiz to check your understanding

1. Why did Cavendish get credit for finding **G**?

    a. He was the first person to measure **G**

    b. It was a mistake, since Newton really discovered **G**

    c. Because others used his experiment to find **G**

2. For what is the mirror used in the Cavendish experiment?

    a. To reflect any light that may disturb the experiment

    b. To measure the angle and oscillation period

    c. There is no mirror used in the experiment

3. Why do the balls oscillate around the equilibrium point?

    a. The value of **G** varies, depending on the angle

    b. Inertia causes them to go past the point and the twisted wire acts as a spring

    c. No one has been able to figure that out yet

Answers

1c, 2b, 3b

10.3 Influence of Gravitation in the Universe

Starting with churning clouds of molecular gases after the Big Bang, gravitation—or the mutual attraction among objects of matter—influenced the formation of the galaxies, stars and planets within the Universe.

The gravitational force is also responsible to the shape of the stars and planets, as well as many other objects in space. The initial motion of all the material in the Universe was later affected by gravitation, resulting in the rotational motion of the galaxies and in the orbits of the planets on our Solar System.

Formation of galaxies, stars and planets

Scientists believe that at about 300,000 years after the Big Bang occurred, the Universe consisted of molecular clouds of ionized gases. Gravitational forces caused clumping of this churning matter until at about 500 million years later, when large, distinct quantities of matter gathered as the beginning of galaxies.

From about 1 billion years after the Big Bang, gravitational forces pulled matter within the galaxies together, forming stars. Because of the large amount of matter in a star, most of it never cooled off. Thermonuclear reactions have kept the stars at extremely high temperature over the eons.

The galaxies and stars continued to evolve until around 9 billion years, when our Solar System was formed along with its planets in motion around the Sun.

Typical spiral-shaped galaxy

Shape of stars and planets

The gravitational attraction during the formation of stars and planets caused them to take on spherical shapes, which is the most efficient shape for evenly distributing the gravitational force among the object's mass.

However, rotation of these bodies on their axes as they cooled often resulted in a bulge at their equators due to centrifugal force. The equatorial radius of the Earth is 6,378.1 km, while its polar radius is 6,356.8 km. Since the Sun revolves so slowly and its mass is so large, the bulge at the Sun's equation is negligible compared to its overall size.

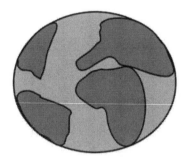

Earth is slightly wider at equator than between poles

Since galaxies are so spread out and consist of gases and stars, the gravitational forces have not been sufficient to pull most of them into even close to a spherical shape.

Smaller objects, such as asteroids and meteors were typically not formed from solidifying liquid and thus are often not spherical in shape.

Keeping objects in orbit

Besides causing quantities of matter to gather together to form suns, planets, moons and other space objects, gravity also caused these moving bodies to go into orbit around each other.

Conditions for orbiting

If the paths of two speeding objects in space intersect, they will collide. The craters you can see on the surface of the Moon are from objects smashing into it. On a larger scale, suns have even collided.

However, if their paths of motion simply go near each other, and if the masses, distances and velocities are just right, the smaller object can go into orbit around the larger object.

(See chapter *12.4 Circular Planetary Orbits* for more information.)

In some cases where the objects are close to the same size, they can rotate around each other. Astronomers have seen a number or double or binary stars that are in orbit around each other.

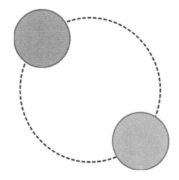

Binary stars orbiting each other

Moon orbits the Earth

Gravitation keeps the Moon in orbit around the Earth.

Since the *Law of Inertia* states that objects in motion tend to go in a straight line, the force of gravitation from the Earth on the Moon prevents it from continuing in a straight path.

It is somewhat like the effect of a string on a weight that you swing around you. Once you let go, the weight no longer goes in a circular path but instead flies out away from you.

Likewise, the Earth and other planets are in orbit around the Sun, and the Sun is in orbit around the center of the Milky Way galaxy.

Summary

After the Big Bang, gravitation influenced the formation of the galaxies, stars and planets within the Universe. Gravitational force is also responsible to the shape of stars and planets, as well as many other objects in space.

The initial motion of all the material in the Universe was later affected by gravitation and resulted in the rotation in galaxies and revolution of planets on our Solar System.

Mini-quiz to check your understanding

1. How were galaxies and stars formed?

    a. They were created when the Big Bang occurred

    b. Gravitation separated objects into distinct galaxies and stars

    c. The force of gravitation clumped masses of material together

2. What causes the bulge at the Earth's equator?

    a. Gravitation from the Sun causes the bulge

    b. The rotation of the Earth on its axis creates an outward force at the equator

    c. The extreme heat at the equator causes material to expand

3. Why do stars revolve around the center of a galaxy?

    a. Gravitation stabilized the initial motion of the stars into orbits

    b. Scientists think it has something to do with gravity

    c. Otherwise they would fly off into space

Answers

1c, 2b, 3a

10.4 Gravitation Causes Tides on Earth

Tides are periodic rise and fall of sea levels, as seen in a specific location on the shore. They are caused by the gravitational forces from the Moon and Sun that attract the ocean water toward them and away from other areas in the ocean.

The rotation of the Earth and the position of the Moon cause the level of the tide to change in a given location. There are two high and low tides each day.

Although you would think the rise in water would only occur on the side toward the Moon and Sun, high tides actually occur on opposite sides of the Earth, caused by a gravitational differential.

The orientation of the Moon and Sun with respect to the Earth determine when the highest and lowest tides occur, as well as when the moderate tides occur.

At the times of the month when the Moon and Sun are aligned, their combined gravitational pull cause the highest tides. The lowest tides are seen at locations on Earth at right angles to the alignment of the Moon and Sun.

Gravitation and tides

If you live near the ocean, you have probably seen the rise and fall of the sea level that happens twice a day. When the sea level is above normal, it is called the high tide. Similarly, low tide is when the sea level on the shore is below normal.

Gravitation from Moon

The gravitational pull on the water from the Moon is the primary cause of the rising tide. Gravitation from the Sun also can contribute to the height of the tide. Centrifugal force on the water from the Earth's rotation also provides a small contribution to the tides.

The gravitational attraction between the Earth and the Moon is F = $1.99*10^{20}$ N (See the chapter on *Gravitational Force Between Two Objects* for the calculations). That force is sufficient to slightly distort the solid surface of both objects toward each other.

Water level rises

Since shape of a body of water can easily be changed, the force from the Moon pulls the ocean toward it a maximum of about one meter.

In some areas where the shore inclination is shallow, a one meter change in sea level can result in even a 10 meter or 40 foot rise in the tide along the shoreline.

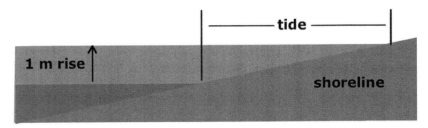

1 meter rise results in several meter rise in tide

High tide every 12 hours 25 minutes

Since the Earth rotates on its axis, the Moon appears to orbit the Earth and is over head every 24 hours and 50 minutes. The extra 50 minutes is a result of the Moon's 27 day actual orbit around the Earth.

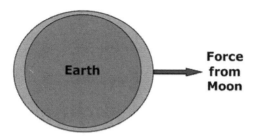

Force from Moon pulls ocean toward it

Although the Moon is overhead every 24 hours and 50 minutes, the high tide comes every 12 hours and 25 minutes. One high tide corresponds to when the Moon is overhead and the other high tide is when the Moon is on the opposite side of the Earth.

Cause of tides on both sides

Since the tides are primarily caused by the gravitation of the Moon acting on the oceans and pulling the surface of the water toward the Moon, you would think the shape of the oceans would be pulled toward the Moon, as opposed to having a high tide on both sides of the Earth. In fact, the configuration seems counter-intuitive.

Simple explanation

A simple explanation for the double tides is that normally a fluid or liquid in space will take on a spherical shape. When you pull or apply a force on one side, the sphere elongates into an oval shape.

Thus, when the Moon pulls the water toward it, the action causes a high tide or bulge on the side of the Earth facing the Moon. But also, the Moon is pulling on the Earth and causes it to move slightly toward it and away from the ocean on the opposite side. This results in the high tide on the side away from the Moon.

Although this explanation is somewhat correct, it really isn't very satisfying.

Theory of the tidal configuration

A more sophisticated explanation is the theory of the tidal configuration which states that the various parts of the Earth's ocean are attracted toward the Moon, according to their separation from the Moon, as well as the angle to the Moon's center. This is also called a gravitational differential.

The force of attraction of the water on the side of the Earth that is closer to the Moon is greater than that on the far side of the Earth.

This is represented in the illustration below by the force-line arrows or vectors.

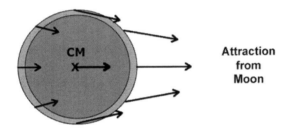

Moon attracts ocean and Earth toward it

But also, the Moon is attracting the mass of the Earth toward it. This can be approximated by considering the mass of the Earth concentrated at its center of mass (CM). This approximation is explained in the *Universal Gravitation Equation* chapter.

The heavy vector represents the attraction of the Earth's mass toward the Moon.

If you subtract the force of attraction on the Earth's center of mass from each of the vectors or force lines to the Moon, the resulting forces on the ocean water are toward and away from the Moon on the ends and moving inward on the sides.

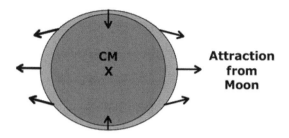

Subtraction of vectors results in double bulge

Tides and orientation of Moon and Sun

Although the gravitational pull from the Moon is the major factor in the creation of the tides, gravitation from the Sun also affects the height of the tide.

When the Sun and the Moon are aligned on the same side of the Earth, it is called a New Moon. With this configuration, the gravitational forces combine and cause a very high tide known as a *spring tide*.

The name has nothing to do with the season and actually occurs slightly after the Moon is overhead, due to the inertia of the ocean and the rotation of the Earth.

When the Sun and Moon are on opposite sides of the Earth, each contributes a pull on the water, resulting in another spring tide. The two spring tides occur two weeks apart.

Alignment of Sun and Moon for spring tides

When the Moon is located at a right angle to the Sun with respect to the Earth, it is called the first quarter or third quarter Moon. In such a case, the difference between the high tide and low tide is much smaller, since the gravitational forces cancel each other. These low tides are called *neap tides*.

Since the orbit of the Moon around the Earth is elliptical, once every 1.5 years the Moon is closest to the Earth. This situation results in an unusually high tide called the *proxigean spring tide*.

Summary

Tides are periodic rise and fall of sea levels that are caused by the gravitational forces of the Moon and Sun on the oceans. There are two high and low tides each day.

High tides occur on opposite sides of the Earth, as do low tides, according to the theory of the tidal configuration. The orientation of the Moon and Sun with respect to the Earth determine when the highest and lowest tides occur, as well as when the moderate tides occur.

Mini-quiz to check your understanding

1. Why does a high tide seem to go further up the shore in some places?

    a. Those places are closer to the Moon

    b. It is an optical illusion

    c. It depends on the inclination of the shoreline

2. What happens when the Moon's gravitation pulls on the ocean?

    a. It causes bulges in the shape of the ocean on both sides of the Earth

    b. It pushes the Earth away from the water

    c. Nothing, unless the Sun is also involved

3. When do spring tides occur?

    a. Only in the spring

    b. Every two weeks

    c. Twice a day

Answers

1c, 2a, 3b

Part 11: Center of Mass

Gravitational measurements are often made with respect to the center of mass (CM) between two objects. Factors about CM are explained in this part.

Part 11 chapters

Chapters in Part 11 include:

11.1 Overview of Gravitation and Center of Mass

This chapter gives an overview of what the CM is and how you can calculate it, the relationship of motion of the objects and how velocity with respect to the CM can be broken into components..

11.2 Center of Mass Definitions

The CM of a sphere and the CM between two spheres are defined. The chapter also discusses some special cases for two spheres.

11.3 Center of Mass Location and Motion

This chapter derives the locations of the objects with respect to the CM between them and shows the relationships of their velocities and accelerations.

11.4 Relative Motion and Center of Mass

The motion relative to an outside observer, with respect to the center of mass and relative to one of the objects are explained.

11.5 Center of Mass Motion Components

This chapter explains how are the motion vectors broken into components, as well as the results of the radial and tangential components of motion.

11.6 Center of Mass and Radial Gravitational Motion

The results of gravitational force attracting two objects toward each other along the radial axis through the center of mass (CM) between them is explained

11.7 Center of Mass and Tangential Gravitational Motion

This chapter shows what happens when objects are moved with different tangential velocities.

11.1 Overview of Gravitation and Center of Mass

The center of mass (CM) of two objects is the weighted average or mean location between their individual centers of mass. It is also called the barycenter. The separation of the objects from the CM is a function of their masses. You can define their locations on a coordinate line and show the relationship of the locations, velocities and accelerations with respect to the CM.

The importance of the CM is that the motion of the two objects is tied to gravitational attraction between them, such that the objects move in opposite directions with respect to that CM. The motion of the objects can be broken into radial and tangential components with respect to the CM.

Radial motion is affected by gravitation, while tangential motion is independent of gravitation. Assuming there are no outside forces acting on them, two objects moving in space will follow curved paths past the CM, go into orbit around the CM or meet each other at the CM.

Definition

A uniform spherical object has its center of mass (CM) at its geometric center. In gravitational calculations, you can assume each object has its mass concentrated at that point.

Of greater interest is the CM between two spheres. Its location is a point that is a ratio of the separations and masses of the objects:

$$mR_m = MR_M$$

$$R = R_m + R_M$$

where

- **m** and **M** are the masses of the two objects
- R_m is the separation between mass **m** and the CM
- R_M is the separation between mass **M** and the CM
- **R** is the separation between masses **m** and **M** as measured from the CM of each sphere

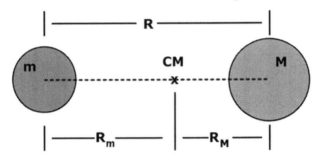

CM between two uniform spheres

(See chapter *11.2 Center of Mass Definitions* for more information.)

Location and motion

The location of the objects on the coordinate line is defined as:

$$r_{CM} = (mr_m + Mr_M)/(m + M)$$

where

- r_{CM} is the location of the CM
- r_m and r_M are the locations of the objects

When the CM is at the zero-point of the coordinate system the objects are on opposite sides of this reference:

$$mr_m = -Mr_M$$

When the CM is considered the reference point, the motions of the objects are in opposite directions with each other and maintain a velocity relationship according to their masses:

$$mv_m = -Mv_M$$

(See chapter *11.3 Center of Mass Location and Motion* for more information.)

However, it is possible to view the objects with respect to an outside reference or relative to one of the objects. In those situations, the CM may be moving and the mass/velocity relationship would not hold.

(See chapter *11.4 Relative Motion and Center of Mass* for more information.)

Components

The motion of objects can be broken into radial and tangential components.

(See chapter *11.5 Center of Mass Motion Components* for more information.)

Only the radial component is affected by gravitation. If there is an initial velocity away from the CM, the objects may move away from each other until they reach a maximum displacement, at which time they reverse directions and move toward each other. If the velocity is sufficient, the objects may escape and fly off into space.

(See chapter *11.6 Center of Mass and Radial Gravitational Motion* for more information.)

The tangential component affects whether the objects will collide, go into orbit or also escape into space.

(See chapter *11.7 Center of Mass and Tangential Gravitational Motion* for more information.)

When the mass **M** is much greater than **m**, the CM is near the center of the object of mass **M**. This greatly simplifies the orbital and escape velocity equations.

Summary

The center of mass (CM) between two objects is the weighted average or mean location of their individual centers of mass. Separation of the objects from the CM is a function of their masses.

Two objects mirror the motion of the other object with respect to the CM. They then maintain the relationship with the ratio of the masses. When viewed with respect to an outside observer, the ratio of the masses may not hold.

Motion can be broken into radial and tangential components. Radial motion is affected by gravitation, while tangential motion is independent of gravitation.

Assuming there are no outside forces acting on them, two objects moving in space will follow curved paths past the CM, go into orbit around the CM or meet each other at the CM.

Mini-quiz to check your understanding

1. What determines the separation ratio of two objects from the center of mass?

   a. It is an arbitrary distance

   b. The masses of the objects

   c. The gravitational constant

2. How do the two objects move with respect to the CM?

   a. They always move in opposite directions

   b. They always move toward the CM

   c. They remain stationary with respect to the CM

3. Why is the tangential velocity component independent of gravitation?

   a. It has no mass

   b. It is in the opposite direction of gravitation

   c. It is perpendicular to the gravitational force

Answers

1b, 2a, 3c

11.2 Center of Mass Definitions

The center of mass (CM) of an object is a point that is the average or mean location of its mass, as if all the mass of the object was concentrated at that point. A uniform sphere has its center of mass at its geometric center.

The CM is sometimes called the *barycenter*.

The CM of a group of objects is point that is the mean location of their individual centers of mass. In our gravitational studies, we are only considering the CM between two objects.

You can find the center of mass between two spheres through a simple ratio formula.

When one object is much larger than the other, the CM may actually be within the larger object.

> **Note**: Some textbooks confuse center of mass with center of gravity (CG), which is related to the effect of gravity on an object, while center of mass concerns mass distribution.
>
> Although CG is often at the same location as the CM, they are completely different concepts.
>
> (See chapter *6.9 Center of Gravity* for more information)

Center of mass of a sphere

The center of mass (CM) of an object is the weighted average of the mass distribution of the body.

In the case of a sphere with the material uniformly distributed, the CM is the geometric center of the object.

Center of mass of sphere is at its geometric center

Approximate center for Earth

Although objects such as the Earth are not exact spheres and do not have their mass uniformly distributed, the variations are small enough to neglect, such that you can consider the CM to be at the geometric center.

CM used in gravitation equation

The Universal Gravitation Equation considers the total mass of a sphere as concentrated at its CM. This assumption simplifies the calculation of the force between two objects, avoiding complex Calculus integration over all particles of the objects.

> (See chapter *8.3 Universal Gravitation Equation* for more information.)

CM between two spheres

In calculating the CM between two spheres—such as between the Earth and the Moon—you can assume each has its mass concentrated at its geometric center.

The center of mass between the spheres is then a point that is a ratio of the separations and masses of the objects:

$$mR_m = MR_M$$

$$R = R_m + R_M$$

where

- **m** and **M** are the masses of the two objects
- R_m is the separation between mass **m** and the CM
- R_M is the separation between mass **M** and the CM
- **R** is the separation between masses **m** and **M** as measured from the CM of each sphere

$mR_m = MR_M$ can also be stated as the inverse ratio of the masses:

$$R_m/R_M = M/m$$

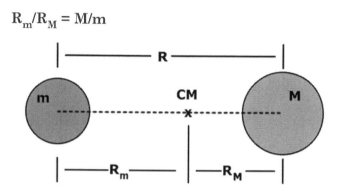

CM between two uniform spheres

Solve for R_m and R_M

If you solve the equations for R_m, you get:

$$R_m = MR_M/m$$

and

$$R_m = R - R_M$$

Combine the equations and solve for R_M:

$$MR_M/m = R - R_M$$

$$MR_M = mR - mR_M$$

Rearrange items:

$$MR_M + mR_M = mR$$

$$R_M (M + m) = mR$$

Thus:

$$R_M = mR/(M + m)$$

Likewise:

$$R_m = MR/(M + m)$$

Special cases

Special cases of the CM between two objects include equal-sized spheres and when one sphere much larger than the other.

Equal sized spheres

For two objects of equal mass, the CM is the point midway between the line joining their centers. Start with the equation:

$$R_m = MR/(M + m)$$

Since:

$$m = M$$

$$R_m = MR/2M$$

Thus:

$$R_m = R/2$$

The illustration shows the relationship.

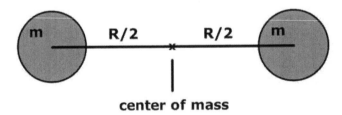

center of mass

Center of mass is at the midpoint for equal objects

CM when one sphere much larger

If one sphere is much larger than the other, the center of mass may even be within the larger object.

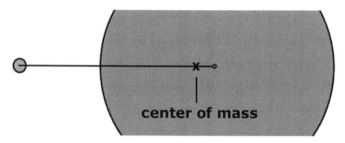

Center of mass can be inside much larger object

Earth and Moon example

A good example concerns the CM between the Earth and the Moon.

The mass of the Earth is $M = 5.974*10^{24}$ kg

The mass of the Moon is $m = 7.348*10^{22}$ kg

$(M + m) = 5.974*10^{24}$ kg $+ 0.073*10^{24}$ kg $= 6.047*10^{24}$ kg

The separation between the Earth and the Moon is $R = 384,403$ km, which equals $3.844*10^5$ km.

Substitute in values to find the separation between the Earth and the CM, R_M:

$R_M = mR/(M + m)$

$R_M = 7.348*10^{22}*3.844*10^5/6.047*10^{24}$ km

$R_M = 4671$ km

Since the radius of the Earth is about 6685 km and the CM between the Earth and Moon is about 4671 km from the center of the Earth, the CM between the Earth and Moon is at about 2014 km below the Earth's surface.

Summary

The center of mass of an object is the average or mean location of its mass. A uniform sphere has its center of mass at its geometric center. The CM of a group of objects is point that is the mean location of their individual centers of mass.

You can find the center of mass between two spheres through a simple ratio formula:

$$mR_m = MR_M$$

$$R = R_m + R_M$$

When one object is much larger than the other, the CM may actually be within the larger object.

Mini-quiz to check your understanding

1. If a spherical object had a greater density of mass on one side, how would that affect its CM?

    a. The CM would be shifted toward the more dense area

    b. The CM would remain at the geometric center, since the object is a sphere

    c. The CM would be shifted toward the less dense area

2. For two spheres, if the **M = 3m**, what would **R_m** equal?

    a. 1/3 of **R_m**

    b. 3 times **R_m**

    c. 3 plus **R_m**

3. Where is the CM between Earth and a space station in orbit around the Earth?

    a. Close to the center of the space station

    b. Halfway between the Earth and the space station

    c. Close to the center of the Earth

Answers

1a, 2b, 3c

11.3 Center of Mass Location and Motion

The separations of two objects from the center of mass (CM) between them are defined by their masses. This information can be used to show the location of the objects and CM on a coordinate line.

You can then establish the velocities of the objects and the CM from the equation of the CM location. Then you can determine their accelerations, provided the two objects are part of a closed system, with no outside forces acting on them.

Typically, you set the position of the CM at the zero-point on the coordinate axis. Velocities are then with respect to this fixed CM location. The only accelerations are due to gravitation along the radial axis.

Locations

In order to establish the relationship between the locations of the objects and the location of the center of mass (CM), consider the points r_m, r_{CM} and r_M along a coordinate line. Relationships are:

$$R_m = r_{CM} - r_m$$

$$R_M = r_M - r_{CM}$$

$$mR_m = MR_M$$

where

- r_{CM} is the position of the CM on the coordinate line
- r_m and r_M are the positions of the objects
- m and M are the masses of the objects

- R_m and R_M are the respective separations from the CM

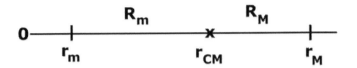

Locations of objects on coordinate line

Location of CM position

You can determine the equation for the CM position r_{CM}:

$$mR_m = MR_M$$

$$m(r_{CM} - r_m) = M(r_M - r_{CM})$$

$$mr_{CM} - mr_m = Mr_M - Mr_{CM}$$

$$r_{CM}(m + M) = mr_m + Mr_M$$

Therefore the general location relationship is:

$$r_{CM} = (mr_m + Mr_M)/(m + M)$$

Locations relative to CM

The general location relationship shows the position of the objects and CM with respect to the zero-point on the coordinate line. In order to locate the objects with respect to the CM, you need to set its location to the zero position on the coordinate line:

$$r_{CM} = 0$$

$$(mr_m + Mr_M)/(m + M) = 0$$

$$mr_m = -Mr_M$$

This means that the locations of the objects are on opposite sides of the zero-point and are a function of their masses.

Velocities

Although you can view the motion of two objects with respect to some outside reference point, our main interest is the motion with respect to the CM, since that is where the gravitational forces are focused.

Velocity is defined as a change in position in a specific direction with respect to an increment of time. Take the derivative with respect to time of the location values to get:

$$v_{CM} = 0$$

$$m(dr_m/dt) = -M(dr_M/dt)$$

$$mv_m = -Mv_M$$

where

- **dr** is the derivative or small change in position **r**
- **dr/dt** is the derivative of **r** with respect to an increment of time
- v_{CM} is the velocity the r_{CM} position in a specific direction
- v_m and v_M are the velocities of the objects with respect to r_{CM}

This means that the velocities move in opposite directions and mirror each other, as seen from the CM.

With respect to external reference

As a point of interest, finding the velocity relationship with respect to an external reference is:

$$v_{CM} = (mv_m + Mv_M)/(m + M)$$

An example of this is if r_{CM} would move perpendicular to the coordinate line, then r_m and r_M would also move in the same direction, as seen with respect to an outside observer.

Accelerations

Acceleration is the change in velocity with respect to time. We are considering the two objects as part of closed system, such that there are no outside forces acting on the objects, except for the gravitational forces between them. In such as case, the only acceleration of the objects is toward the CM. In other words:

$$a_{CM} = 0$$

$$ma_{Rm} = -Ma_{RM}$$

where

- a_{Rm} is radial acceleration of mass **m** toward the CM
- a_{RM} is radial acceleration of mass **M** toward the CM

This means that the radial acceleration vectors are in opposite directions, when viewed from the CM.

Summary

The location of the two objects and the CM between them, with respect to the zero-point of the coordinate line, follows the relationship:

$$r_{CM} = (mr_m + Mr_M)/(m + M)$$

Typically, you set the position of the CM at the zero-point in the coordinate system, so the relationship between locations is:

$$mr_m = -Mr_M$$

Likewise, when the velocity of the CM is zero, the velocities of the objects with respect to the CM are:

$$mv_m = -Mv_M$$

Acceleration is only in the radial direction, due to gravitation:

$$ma_{Rm} = -Ma_{RM}$$

When observed with relative to the CM, the motions of the objects mirror each other.

Mini-quiz to check your understanding

1. How is the relationship for locations of the objects determined?

   a. Be setting location points on the coordinate line

   b. By comparing their masses

   c. By setting the CM to zero

2. How do you determine the velocity of an object from its position?

   a. They are the same items

   b. Velocity is a small change in position with respect to an increment in time

   c. A change in position indicates the velocity is zero

3. Why is the only acceleration due to gravitation?

   a. The motion of the objects compensate for other accelerations

   b. It is a fact that no one understands

   c. Gravitation is the only force acting on the object

Answers

1a, 2b, 3c

11.4 Relative Motion and Center of Mass

The motion of two objects in space is relative to some defined point of reference. You may observe the motion with respect to your own viewpoint or some fixed reference.

However in such a case, you often cannot tell where the center of mass (CM) is actually located.

Ideally, the point of reference is the CM between the two objects. This provides the best results and information about the motion. Another point of reference is that the motion can be with respect to one of the two objects, such as viewing the motion of the Moon with respect to the Earth.

Motion relative to outside observer

When you observe two moving objects in space, your point of view concerning that motion usually is that you are stationary and the objects are moving with respect to you. This is especially evident when both objects are moving in the same direction.

Your viewpoint of the motion of the center of mass (CM) is a function of the velocities of the objects relative to you or to some other stationary object and their masses. The general velocity relationship is:

$$v_{CM} = (mv_m + Mv_M)/(m + M)$$

where

- m and M are the masses of the objects
- v_{CM} is the velocity of the CM
- v_m and v_M are the velocities of the objects

The system of two objects appears to be moving, and it is difficult to tell where their CM is located.

Motion relative to CM

Most often, the motion of two objects in space is illustrated as with respect to a stationary CM. This is the ideal case. For example, when you observe two objects on orbit around their CM—such as with twin stars—you can usually see them orbit the fixed CM between them. Your view is relative to that CM, such that $v_{CM} = 0$.

The general velocity equation becomes:

$$0 = (mv_m + Mv_M)/(m + M)$$

$$mv_m = -Mv_M$$

This means that the velocity vectors are in opposite directions, when viewed from the CM. In this ideal case, it can be said that the motions of the objects mirror each other, according to their respective masses. This "mirroring" can be seen when both objects move apart, move toward each other, go into orbit and fly off into space.

Motion relative to one object

Instead of the two objects moving with respect to you or relative to the CM, the motion of one object is sometimes viewed with respect to the other object, as if it were stationary.

An example of this is when you view the Moon's orbit from the Earth.

If the object of mass **m** is seems to be moving with respect to object **M**, its velocity is the difference in their velocity vectors with respect to the CM:

$$V_{mM} = V_m - V_M$$

where

- V_{mM} is the velocity vector of object **m** with respect to object **M**
- V_m and V_M are the velocity vectors of the two objects with respect to the CM

Since the velocity vectors are in opposite directions or $\mathbf{mv_m} = -\mathbf{Mv_M}$, the magnitude of $\mathbf{v_{mM}}$ is the sum of the magnitudes of the two velocities.

(See chapter *12.2 Orbital Motion Relative to Other Object* for more information.)

Summary

The motion of two objects in space is relative to some fixed reference point. In many cases, you may see the motion with respect to you viewpoint. Ideally, the viewpoint is relative to a stationary CM between the two objects. It is also possible to view the motion with respect to one of the two objects.

Mini-quiz to check your understanding

1. If you see two objects in space moving in the same direction and parallel to each, how is the CM between them moving?

    a. The CM is stationary, since the objects are moving together

    b. The CM rotates on its axis

    c. It is moving along with them in the same direction

2. If an object is moving toward the stationary CM, how does the other object mirror the motion?

    a. It moves away from the CM at the same speed

    b. It also moves toward the CM but in the opposite direction

    c. It depends on how fast the CM is moving

3. Why would you view the motion of one object with respect to the other?

    a. It may be a more convenient perspective

    b. You would never view one object with respect to the other

    c. You always do this when on the Earth

Answers

1c, 2b, 3a

11.5 Center of Mass Components

The motion vectors of two objects in space with respect to the center of mass (CM) between them can be broken into their radial and tangential components. One reason to break the motion into its components is to facilitate calculations, especially since radial motion is affected by the gravitational attraction between the objects, while tangential speed in unaffected by gravitation.

When viewed from the CM, the objects move in the radial direction either both toward the CM or both away from the CM. The objects move in opposite directions tangential to the line through the CM.

The combination of the radial and tangential motion components determine whether the objects will collide, go into orbit or fly off into space.

Motion broken into components

When two objects in space are viewed as moving with respect to the CM between them, their motion can be broken into radial and tangential vectors in the coordinate system, with the CM as the fixed-axis point.

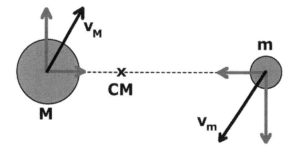

Motion of objects can be broken into components

Gravity and Gravitation

The radial vector components are on the axis through the CM. The tangential components are perpendicular to that axis.

Opposite directions

When viewed with respect to the CM, the velocity vectors follow the ratio:

$$mv_m = -Mv_M$$

This expression simply means that the velocity vectors are in opposite directions, when viewed from the CM. It is not to be confused with an overall direction convention.

Radial component

The radial velocity vector is:

$$v_R = v*cos(\theta)$$

where

- v_R is the radial velocity vector
- v is the velocity of the object
- θ (Greek letter theta) is the angle
- $cos(\theta)$ is the cosine of angle theta

Note: The angle is measured from the radial axis to the velocity vector, in a counterclockwise direction.

When $\theta > 90°$, $cos(\theta)$ is negative. Thus v_R is pointing away from the CM and is also negative.

Note: This follows that $mv_m = -Mv_M$ means the vectors are in opposite directions.

Tangential component

The tangential velocity vector is:

$$v_T = v*sin(\theta)$$

where

- v_T is the tangential velocity vector
- $\sin(\theta)$ is the sine of angle theta

When $\theta > 180°$, $\sin(\theta)$ is negative. Thus v_T is pointing in a counterclockwise direction and is also negative.

Radial vectors

The radial vectors are along the axis between the objects. When viewed with respect to the CM, their motion is either toward the CM or away from the CM. The radial velocities between the objects are indicated as v_{Rm} and v_{RM}.

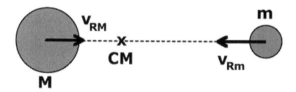

Radial vector components of the two objects

> **Note**: If the objects appear to be moving in the same direction, such that one object appears to be moving toward the CM, while the other is moving away from the CM, the perspective is not with respect to the CM. Changing the point of view, corrects the apparent motion.

Assuming the tangential velocity is zero, the objects will only move on the radial axis, according to the initial direction of their velocity and the effect of gravitation. The addition of tangential motion to the objects can cause then to go into orbit or fly off into space.

Motion away from CM

If the objects are moving away from the CM, they may reach a maximum displacement and then change directions and fall back toward each other and the CM. This is similar to the case of throwing a ball upward and having it return to Earth.

(See chapter *5.1 Overview of Gravity Equations for Objects Projected Upward* for more information.)

If the initial radial velocity in a direction away from the CM is sufficient, the objects can escape from their gravitational attraction.

(See chapter *11.6 Center of Mass and Radial Gravitational Motion* for more information.)

Tangential vectors

A tangential vector is perpendicular to the radial component and the axis between the objects. The tangential velocities of the objects are indicated as v_{Tm} and v_{TM}. The tangential velocities are in opposite directions when viewed relative to the CM.

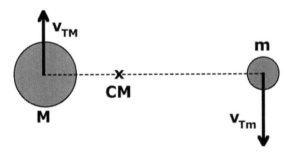

Tangential vector components of the two objects

Since there is always motion in the radial direction, the combination of the tangential and radial velocities will determine whether the objects will move past each other and go off into space, go into orbit around the CM or collide.

(See chapter *11.7 Center of Mass and Tangential Gravitational Motion* for more information.)

Amount of tangential velocities

The combination of the tangential velocities and the gravitational attraction toward the CM, cause the objects to move in curved paths about the CM.

Depending on the tangential velocity, the objects will move in elliptical or circular orbits. If the velocity is great enough, the objects will go into parabola or hyperbolic paths and escape into space.

If there are no tangential velocities, the objects will move inward and collide at the CM.

Summary

The gravitational motion of two objects in space can be viewed with respect to the center of mass (CM) between them.

The motion vector of each object can be broken into its radial and tangential components, with respect to that CM. Since there is always a gravitational attraction between the objects, there is always a radial component to their motion.

The combination of the radial and tangential motion components determine whether the objects will collide, go into orbit or fly off into space.

Mini-quiz to check your understanding

1. What is the relationship between the radial and tangential velocity components?

    a. The radial component is always larger

    b. The components are perpendicular to each other

    c. The components have no real relationship

2. In which direction do the objects move on the axis between them as viewed from the CM?

    a. Toward each other or away from each other

    b. Objects only move toward each other because of gravitation

    c. It depends on the magnitude of the tangential velocity

3. What is necessary for objects to orbit the CM?

    a. Their tangential velocities must be sufficient for orbits to occur

    b. The objects must have the same mass

    c. Two objects always orbit their CM

Answers

1b, 2a, 3a

11.6 Center of Mass and Radial Gravitational Motion

The motion of two objects in space can be broken into perpendicular radial and tangential vector components with respect to the center of mass (CM) between them.

The radial components are along the axis through the CM and are affected by the gravitational force between the two objects, as well as any initial velocities.

Gravitational attraction will normally pull the objects toward each other along the radial axis.

However, if there is an outward initial velocity, the objects may be moving away from the CM, where they can reach a maximum displacement and return toward each other or escape the gravitational pull and fly off into space.

The amount of the tangential components determine whether the objects will collide at their CM, go into orbit around each other or simply pass by and fly off into space.

If the objects seem to be moving in the same direction, the viewpoint is with respect to an outside observer. This view can be transformed to be with respect to the stationary CM.

Both objects moving toward CM

Two objects in space may be moving with the radial components of their velocity vectors in a direction toward each other, due to the gravitational pull between them, as well as initial velocities toward the CM.

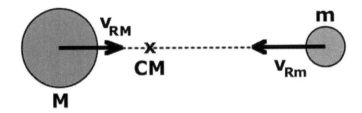

Objects moving toward CM

Velocity relationship

The relationship between the velocities is:

$$mv_{Rm} = -Mv_{RM}$$

where

- **m** and **M** are the masses of the objects
- v_{Rm} and v_{RM} are the radial velocities

Note: The negative sign indicates the velocity vectors are in opposite directions.

Beyond velocity relationship

If the velocities do not appear to follow the above relationship, your point-of-view is not with respect to the CM. Instead, it may be with respect to some other point-of-view.

Effect of tangential velocities

The amount of the tangential component determines whether the objects will collide at their CM, go into orbit around each other or simply pass by and fly off into space.

Acceleration relationship

The relationship between the accelerations of the objects with respect to the CM is:

$$ma_{Rm} = -Ma_{RM}$$

These accelerations also are related to the gravitational force. Consider:

$$F = GMm/R^2$$

where

- **F** is the force of attraction between two objects
- **G** is the Universal Gravitational Constant
- **R** is the separation between the centers of the objects

Compare with the force-acceleration relationship for mass **m**:

$$F_{Rm} = ma_{Rm}$$

$$a_{Rm} = GM/R^2$$

Likewise,

$$F_{RM} = Ma_{RM}$$

$$a_{RM} = Gm/R^2$$

Although the accelerations are in opposite directions toward the CM, their magnitudes are related to the mass of the attracting object and the separation of the objects.

Objects moving away from each other

For the objects to be moving away from each other, there must have been some initial impetus or force applied to give them their velocities. That force is no longer applied when we examine the motion of the objects. although they still have their initial outward velocities.

Objects moving away from each other

Depending on their initial velocities, the objects may reach some maximum displacement and reverse directions to move toward the CM. This is similar to the effect of throwing a ball into the air and having it return to Earth.

(See chapter *5.2 Velocity Equations for Objects Projected Upward* for more information.)

If the velocities are sufficient, the objects may escape the gravitational pull of each other.

Example of expansion of Universe

A good example of various sized objects moving away from the center of mass between them is the expansion of the Universe, where the galaxies have been measured as moving away from some center point.

Speculation on the expansion of the Universe after the Big Bang occurrence is whether the expansion is at a rate where the objects will move outward forever or whether the galaxies will reach a maximum displacement and reverse directions back toward the CM of the Universe.

Measurements on the rate of expansion have determined that "something" is affecting it—perhaps dark matter.

(See chapter *8.6 Effect of Dark Matter and Dark Energy on Gravitation* for more information.)

Example of escape velocity by rocket

When one object is much larger than the other, the CM is close to the geometric center of the larger object. An example of this is the comparison of a rocket and the much larger Earth.

If the rocket is sent upward at a sufficient velocity, it can escape the gravitational pull between it and the Earth, such that it will go off into space.

(See chapter *13.1 Overview of Gravitational Escape Velocity* for more information.)

Effect of tangential component

In cases where the velocities are less than the escape velocity, the tangential velocities will determine whether the objects fall back into each other or go into orbit around the CM.

Objects moving in same direction

You may observe two objects moving in the same radial direction. This means you are observing the object with respect to some other point of reference and not relative to the CM. When the objects are moving in the same direction, the CM is moving along with the objects. In such a case, the $\mathbf{mv_{Rm}} = -\mathbf{Mv_{RM}}$ ratio does not hold.

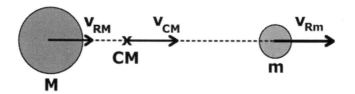

Objects moving in the same direction

If the viewpoint is changed to be with respect to the CM, the CM will appear stationary and the objects will be either both moving toward the CM or away from that point.

Summary

The radial component of the motion of two objects in space is along the axis through the CM.

The gravitational attraction normally pulls the objects toward each other along the radial axis. But if there is an outward initial velocity, the objects may be both moving away from the CM, where they can reach a maximum displacement and return toward each other or escape the gravitational pull and fly off into space.

The tangential motion component affects whether the objects will collide at their CM, go into orbit around each other or simply pass by and fly off into space.

If the objects seem to be moving in the same direction, the viewpoint is with respect to an outside observer. This view can be transformed to be with respect to a stationary CM.

Mini-quiz to check your understanding

1. What does $mv_{Rm} = -Mv_{RM}$ indicate?

   a. It means that v_{RM} is less than zero

   b. The equation is incomplete

   c. The negative sign indicates that v_{RM} is in the opposite direction of v_{Rm}

2. What can happen when the objects are moving away from each other with respect to the CM?

   a. They start to go into orbit around each other

   b. They may reach a maximum displacement and reverse directions

   c. One object will stop while the other continues on

3. What happens to the CM if it appears to you that the objects are moving in the same direction?

   a. The CM also moves in that direction to an outside observe

   b. The CM stays where it was originally

   c. The CM moves in an opposite direction of the objects

Answers

1c, 2b, 3a

11.7 Center of Mass and Tangential Gravitational Motion

Tangential motion of two objects in space, with respect to the center of mass (CM) between them, is perpendicular to the radial line between the objects and through the CM. The objects move in opposite tangential directions with respect to the CM.

While the radial motion components are a function of the gravitational force between the objects, tangential velocities are not affected by gravitation.

The tangential velocities determine whether the objects will collide, go into orbit or fly off into space. When there are additional radial velocities away from or toward the CM, the paths of the objects can be distorted.

Objects move in opposite directions

When the tangential components of the motion of two objects in space viewed with respect to the center of mass (CM), they are always in opposite directions, with the ratio of the velocities as:

$$m v_{Tm} = -M v_{TM}$$

where

- m and M are the masses of the objects
- v_{Tm} and v_{TM} are the tangential velocities of the objects

The negative sign indicates the velocity vectors are pointing in opposite directions. In other words, the objects are moving in opposite directions.

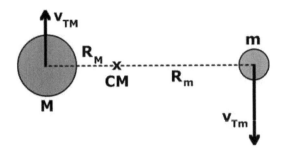

Factors in tangential motion

Ratio of speeds

Since speed is independent of direction:

$$ms_{Tm} = Ms_{TM}$$

$$s_{Tm}/s_{TM} = M/m$$

where s_{Tm} and s_{TM} are the speeds or magnitudes of velocities.

Other viewpoint

If the velocities do not appear to be in the given ratio or if both objects are moving in the same direction, the viewpoint is with respect to an outside observer and not relative to the CM. In these cases, though, the CM may appear to be moving in order to maintain the correct ratio.

(See chapter *11.4 Relative Motion and Center of Mass* for more information.)

Results determined by velocity

Although the objects are moving perpendicular to the axis between them, they are still attracted toward each other by their gravitational force.

The combination of the tangential velocities and the gravitational attraction toward the CM along the radial direction causes the objects to move in curved paths. The result of the curved motion is an outward centrifugal forces on each of the objects.

Possible paths of the objects are that they:

Collide

Go into orbit

Small elliptical orbits

Circular orbits

Large elliptical orbits

Fly off into space

Parabolic paths

Hyperbolic paths

(See chapter *12.6 Effect of Velocity on Orbital Motion* for more information.)

Collide

When the tangential velocities are zero ($v_{Tm} = v_{TM} = 0$), the only motion is toward the CM, where the objects collide.

Go into small elliptical orbits

When the tangential velocities are less than required for a circular orbit, the objects will follow shallow elliptical paths. Depending on the physical size of the objects, they may either collide or go into orbit around each other.

Go into circular orbits

At specific tangential velocities for given masses and separations, the two objects will rotate about the CM between them in circular orbits. The centrifugal forces of the objects keep them in this stable orbit.

> **Note**: Assume that there is no extra radial velocity, either toward or away from the CM that would affect the orbit.

The tangential velocities of the objects with respect to the CM are:

$$v_{Tm} = \sqrt{[GM^2/R(M + m)]}$$

$$v_{TM} = \sqrt{[Gm^2/R(M + m)]}$$

where

- v_{Tm} and v_{TM} are the tangential velocities
- **G** is the Universal Gravitational Constant
- **M** and **m** are the masses of the two objects
- **R** is the separation between the objects, as measured from their centers of mass

(See chapter *12.1 Derivation of Circular Orbits Around Center of Mass* for more information.)

Go into elliptical orbits

When the velocities are greater than those required for circular orbits but less than the escape velocity, the objects will go into elliptical orbits around the CM. The velocity range for mass **m** is:

$$\sqrt{[GM^2/R(M + m)]} < v_{Tm} < \sqrt{[2GM^2/R(M + m)]}$$

Follow parabolic paths

If the velocities are at the escape velocity, the objects will take parabolic paths into space. The velocity for mass **m** is:

$$v_{Tm} = \sqrt{[2GM^2/R(M + m)]}$$

If the velocity is *with respect to the other object*, the resulting escape velocity is:

$$v_T = \sqrt{[2G(M + m)/R]}$$

(See chapter *13.1 Overview of Gravitational Escape Velocity* for more information.)

Follow hyperbolic paths

If the velocities are above the escape velocity, the objects will follow hyperbolic paths and go off into space.

Effect of initial radial velocity

Besides the radial velocity from the gravitational attraction of the two objects, there may be an initial velocity, either toward the CM or away from the CM. In both cases, the paths of the objects are distorted according to the constant radial velocity.

Radial velocity toward CM

When the objects are moving in circular orbits, motion toward the CM is compensated by motion away from the centrifugal force.

However, when the objects are moving toward the CM at some constant velocity, the effect would be to reduce the separation and change the shape of the paths of the objects.

The results could be changing the circular orbits into elliptical or even spirals into the CM.

Radial velocity away from CM

When the objects are moving away from the CM, such that they reach a maximum displacement before falling toward the CM, the effect of the tangential velocity is distorted.

For example, in order to be in a circular orbit, the tangential velocity must create a sufficient centrifugal force to equalize the gravitational force.

However, the motion away from the CM increases the separation, thus requiring a higher tangential velocity for the orbits.

The result would be elliptical shaped orbits until the objects returned to the initial separation needed for circular orbits. It is not easy to visualize, and the math required is complex.

Of course, if the outward radial velocities were at or above the escape velocities, the objects would follow curved paths as they flew into space.

Summary

Tangential motion of two objects in space is perpendicular to the radial line between the objects and in opposite directions with respect to the CM. Tangential velocities are not affected by gravitation.

The tangential velocities determine whether the objects will collide, go into orbit or fly off into space. When there are additional radial velocities, the paths of the objects are distorted.

Mini-quiz to check your understanding

1. What is the ratio of tangential speeds with respect to the CM?

    a. It is $s_{TM}/s_{Tm} = M/m$

    b. It is $v_{Tm}/v_{TM} = M/m$

    c. It is $s_{Tm}/s_{TM} = M/m$

2. What happens when you slightly increase the tangential velocity of a circular orbit?

    a. The orbit becomes elliptical

    b. The object will fly off into space in a radial direction

    c. It becomes a larger circle

3. How does an additional radial component affect the object's motion?

    a. It negates the tangential motion

    b. It can cause the object to move toward or away from the CM

    c. You cannot have an additional radial component

Answers

1c, 2a, 3b

Part 12: Orbital Motion

An object in space may exhibit orbital motion or curved paths with respect to another object due to its velocity and the gravitational force between the objects.

Part 12 chapters

Chapters in Part 12 include:

12.1 Derivation of Circular Orbits Around Center of Mass

This chapter shows that when the gravitational force equals the inertial centrifugal forces of two objects, they will go in circular orbits around their center of mass (CM).

12.2 Orbital Motion Relative to Other Object

One object can appear to orbit the other when the point of view is changed from the CM to the other object.

12.3 Direction Convention for Gravitational Motion

This chapter defines a convention for which direction is positive and which is negative for motion with respect to another object.

12.4 Circular Planetary Orbits

The velocities of the Moon orbiting the Earth, the Earth orbiting the Sun and the planet Jupiter orbiting the Sun are calculated in this chapter.

12.5 Length of Year for Planets in Gravitational Orbit

There are simple equations to calculate the length of a year or single rotation for planets in gravitational orbit. This chapter shows the velocity necessary to be in orbit and gives some examples to verify the equations.

12.6 Effect of Velocity on Orbital Motion

The magnitude of the tangential velocity can determine whether an object will spiral into the other, go into orbit or fly off into space. This chapter gives the equations for each situation.

12.1 Derivation of Circular Orbits Around Center of Mass

Circular orbits of two objects around the center of mass (CM) between them require tangential velocities that equalize the gravitational attraction between the objects.

Tangential velocities tend to keep the objects traveling in a straight line, according to the Law of Inertia.

If gravitation cases an inward deviation from straight-line travel, the result is an outward centrifugal force. By setting the gravitational force equal to the centrifugal forces, you can derive the required tangential velocities for circular orbits.

The orbit equations can be in simplified forms when the masses of the two objects are the same and when the mass of one object is much greater than that of the other.

Factors in determining orbital velocities

The linear tangential velocities required for two objects to be in circular orbits around the CM between them is found by comparing their gravitational force of attraction with the outward centrifugal force for each object.

> **Note**: A linear tangential velocity is a straight line velocity is perpendicular to the axis between the two objects. It is tangent to the curved path and is different from rotational velocity.
>
> (See chapter *11.7 Center of Mass and Tangential Gravitational Motion* for more information.)

Assume no initial radial velocities

When two objects in space are traveling toward the general vicinity of each other, they both have radial and tangential veloci-

---

ties with respect to the center of mass (CM) between them. However, in order to simplify the derivation for circular orbits, we will only look at the case where there are no inward or outward radial velocities and only be concerned about the tangential velocities.

This is similar to the case of Newton's cannonball going into orbit or sending a satellite into orbit around the Earth.

> (See chapter *6.6 Gravity and Newton's Cannon* for more information.)

Since there is no radial motion, separation between the objects remains constant, which is a requirement for circular orbits

Gravitational force of attraction

The gravitational force of attraction between two objects is:

$$F = GMm/R^2$$

where

- **F** is the force of attraction between two objects in newtons (N)
- **G** is the Universal Gravitational Constant= $6.674*10^{-17}$ N-km^2/kg^2
- **M** and **m** are the masses of the two objects in kilograms (kg)
- **R** is the separation in kilometers (km) between the objects, as measured from their centers of mass

Note: Since force is usually stated in newtons, but motion between astronomical bodies is usually stated in km/s, an adjusted value for **G** is used, with N-km^2/kg^2 as the unit instead of N-m^2/kg^2. **G** is also sometimes stated as: $6.674*10^{-20}$ km^3/kg-s^2.

(See chapter *8.3 Universal Gravitation Equation* for more information.)

Separation of objects

Although the objects are orbiting the CM, their separation, **R**, remains constant. The individual separations between the objects and CM are also constant and determined by **R** and their masses:

$$R = R_M + R_m$$

where

- R_M is the separation between the center of object **M** and the CM in km
- R_m is the separation between the center of object **m** and the CM in km

The values of R_M and R_m are according to the equations:

$$R_M = mR/(M + m)$$

$$R_m = MR/(M + m)$$

(See chapter *11.1 Overview of Gravitation and Center of Mass* for more information.)

The factors involved can be seen in the illustration below:

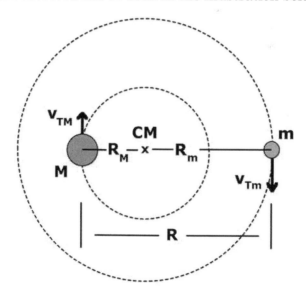

Factors in objects orbiting CM

Note: Although the Earth orbits the Sun in a counterclockwise direction, we usually indicate motion in a clockwise direction.

(See chapter *12.3 Direction Convention for Gravitational Motion* for more information.)

Centrifugal force

The centrifugal inertial force on each object relates to its circle of travel:

$$F_M = Mv_{TM}^2/R_M$$

$$F_m = mv_{Tm}^2/R_m$$

where

- F_M is the centrifugal inertial force on mass **M**
- v_{TM} is the tangential velocity of mass **M**
- F_m is the centrifugal inertial force on mass **m**
- v_{Tm} is the tangential velocity of mass **m**

Note: Centrifugal force is caused by inertia and is not considered a "true" force. It is sometimes called a pseudo- or virtual force.

Substituting $R_M = mR/(M + m)$ and $R_m = MR/(M + m)$ in the above equations gives you:

$$F_M = Mv_{TM}^2(M + m)/mR$$

$$F_m = mv_{Tm}^2(M + m)/MR$$

Solve for individual velocities

Since the centrifugal force equals the gravitational force for a circular orbit, you can solve for the velocity.

Object with mass m

In the case of the object with mass **m**:

$$F_m = F$$

Gravity and Gravitation

Substitute equations:

$$mv_{Tm}^2(M + m)/MR = GMm/R^2$$

Multiply both sides by **MR** and divide by **m**:

$$v_{Tm}^2(M + m) = GM^2/R$$

Divide by **(M + m)**:

$$v_{Tm}^2 = GM^2/R(M + m)$$

Take the square root of both sides:

$$v_{Tm} = \pm\sqrt{[GM^2/R(M + m)]}$$

This means the velocity can be in either direction for a circular orbit. Since direction is not relevant here:

$$v_{Tm} = \sqrt{[GM^2/R(M + m)]} \text{ km/s}$$

Object with mass M

Likewise, for the object of mass **M**:

$$v_{TM} = \sqrt{[Gm^2/R(M + m)]} \text{ km/s}$$

Sizes of objects

The equations for the tangential orbital velocities can be simplified when both objects are the same size, as well as when one object has a much greater mass than the other.

Objects have same mass

There are situations in space where two stars have close to the same mass and orbit the CM between them. Astronomers call them *double stars.*

If the objects are the same mass, then **M = m** and the velocity equation for each becomes:

$$v_{TM} = \sqrt{[Gm^2/R(m + m)]} \text{ km/s}$$

The equation reduces to:

$$v_{TM} = \sqrt{[Gm/2R]} \text{ km/s}$$

Since both objects or stars have the same orbital velocity and the same separation from the CM, they follow the same orbit around the CM.

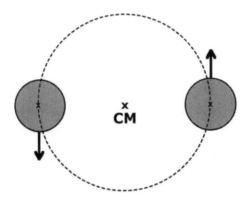

Double stars follow same orbit around CM

One object much larger than other

Another situation often seen in space is when one object is much larger than the other. In this case, the CM between them is almost at the larger object's geometric center.

This results in simplifying the equation for orbital velocity. The small object then seems to orbit the larger object.

For example, the CM between a satellite orbiting the Earth is near the geometric center of the Earth. Likewise, the CM between the Earth and the Sun is near the center of the Sun.

Suppose **M >> m** (**M** is much greater than **m**). Then:

$$M + m \approx M$$

where $\approx$ means "approximately equal to".

Substitute $\mathbf{M + m \approx M}$ into the equation for the velocity of the smaller object:

$$v_{Tm} = \sqrt{[GM^2/R(M + m)]}$$

$$v_{Tm} = \sqrt{[GM^2/RM]}$$

Reducing the equation results in:

$$v_{Tm} = \sqrt{(GM/R)} \text{ km/s}$$

This is the same as the standard equation for orbital velocity of one object around another.

(See chapter *12.2 Orbital Motion Relative to Other Object* for more information.)

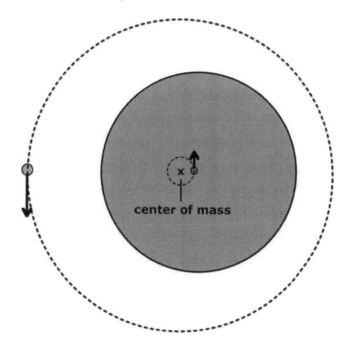

center of mass

Orbits when one object is much larger than other

Summary

When two objects are moving at the correct tangential velocities, they will go in circular orbits around their CM.

The velocity equations are determined by setting the gravitational force equal to the outward centrifugal forces caused by their tangential velocities. The velocity equations are:

$$v_{Tm} = \sqrt{[GM^2/R(M + m)]} \text{ km/s}$$

$$v_{TM} = \sqrt{[Gm^2/R(M + m)]} \text{ km/s}$$

When the mass of each object is the same, the velocity equation is simplified. The same is true when the mass of one object is much greater than that of the other.

Mini-quiz to check your understanding

1. What do you compare to find the circular orbital velocity?

    a. Gravitational force and centrifugal force

    b. Radial and tangential velocities

    c. The mass of the larger versus the mass of the smaller object

2. What is the relationship between tangential velocity for circular orbits and mass?

    a. Velocity is independent of the mass of the objects

    b. The velocity of an object is a function of the mass of the other object

    c. The velocity of an object is only dependent on its mass

3. What appears to happen when one object is much larger than the other?

    a. The larger object appears to orbit the much smaller object

    b. The smaller object appears to orbit the much larger object

    c. The objects remain stationary in space

Answers

1a, 2b, 3b

12.2 Orbital Motion Relative to Other Object

The orbital motion of two objects in space is often seen by an external observer as rotating about the center of mass (CM) or barycenter between them, as if the CM was a fixed axis-point.

However, to an observer on one of the objects, that object is fixed, while the other object appears to be orbiting the "fixed" object.

A good example is how the Moon appears to orbit the Earth. An advantage of considering the orbit of one object around the other is that it simplifies the two orbital equations into one. This viewpoint is more convenient for calculating orbits and even escape velocity.

You can find the relative velocity of the one object as seen from the other by adding the two tangential velocity equations for circular orbits around the CM together and then rearranging the terms. A further simplification is seen when one object has a much greater mass than the other.

The equation for orbital velocity of one object with respect to the other is independent of which object is in orbit.

Tangential velocities of orbiting objects

The method to find the tangential velocity of one object with respect to the other is to start with the tangential velocities of the two objects with respect to the center of mass (CM) between them.

Velocities relative to CM

The tangential velocities of two objects in circular orbits around the center of mass (CM) between them and relative to that CM are in opposite directions, according to the relationship:

$$mv_{Tm} = -Mv_{TM}$$

The requirement for circular orbits around the CM is that the tangential velocities with are:

$$v_{Tm} = \sqrt{[GM^2/R(M + m)]} \text{ km/s}$$

$$v_{TM} = \sqrt{[Gm^2/R(M + m)]} \text{ km/s}$$

where

- v_{Tm} is the tangential velocity of mass **m** in km/s
- v_{TM} is the tangential velocity of mass **M** in km/s
- **G** is the Universal Gravitational Constant= $6.674*10^{-20}$ km^3/kg-s^2
- **m** is the mass of the smaller object in kg
- **M** is the mass of the larger object in kg
- **R** is the separation in kilometers (km) between the objects, as measured from their centers of mass

(See chapter *12.1 Derivation of Circular Orbits Around Center of Mass* for more information.)

Velocity with respect to other object

If you shift the zero-point from the CM to the center of mass **M**, r_M, the tangential velocity of mass **m** is the difference of the tangential velocities with respect to the CM.

$$v_T = v_{Tm} - (-v_{TM})$$

Since the tangential velocities are in opposite directions, the magnitude of v_T is simply the sum of the two speeds and its direction is the same as v_{Tm}. In other words, you can add the velocity equations.

The same logic holds if you consider the velocity of **M** with respect to **m**. In that case, $v_T = v_{TM} - (-v_{Tm})$. In other words, it does not matter which object is considered fixed.

Finding velocity relative to other object

You can find the tangential velocity of one object with respect to the other by adding their tangential velocities with respect to the CM:

$$v_T = \sqrt{[GM^2/R(M + m)]} + \sqrt{[Gm^2/R(M + m)]}$$

Some algebraic manipulation is necessary:

$$v_T = M\sqrt{(G)}/\sqrt{[R(M + m)]} + m\sqrt{(G)}/\sqrt{[R(M + m)]}$$

Combine both fractions over same denominator:

$$v_T = [M\sqrt{(G)} + m\sqrt{(G)}]/\sqrt{[R(M + m)]}$$

$$v_T = [(M + m)\sqrt{(G)}]/\sqrt{[R(M + m)]}$$

Note that $(M + m) = \sqrt{(M + m)^2}$:

$$v_T = \sqrt{[G(M + m)2/R(M + m)]}$$

Thus the tangential velocity for a circular orbit, as seen from the other object is:

$$v_T = \sqrt{[G(M + m)/R]} \text{ km/s}$$

In the case that **M >> m** (M is much greater than **m**), the equation reduces to:

$$v_T = \sqrt{(GM/R)} \text{ km/s}$$

Example of relative motion

An example of this relative motion is how the Moon appears to orbit the Earth. However, from the viewpoint of the Moon, the Earth appears to orbit the Moon at the same velocity.

Moon orbits the Earth

Considering the Earth has mass **M** and the Moon has mass **m**, you can see how the Moon appears to orbit the Earth in a counterclockwise direction:

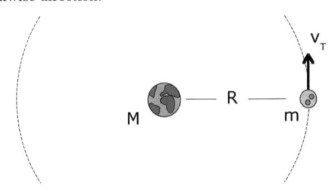

Moon appears to orbit Earth

The CM between the objects is within the Earth's surface. That CM follows the Moon's orbit around the Earth.

Earth orbits the Moon

When astronauts were on the Moon, they saw the Earth orbiting the Moon.

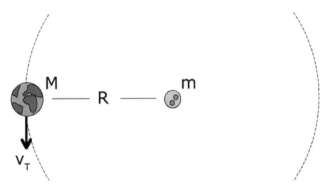

Earth appears to orbit Moon

Summary

The orbital motion of two objects in space is often seen by an external observer as rotating about the center of mass between them. However, to an observer on one of the objects, that object is fixed and the other object appears to be orbiting the "fixed" object.

You can find the relative velocity of the one object as seen from the other by taking the two tangential velocities for circular orbits around the CM and adding them together. The relative velocity is then the sum of the tangential velocities:

$$v_T = \sqrt{[G(M + m)/R]} \text{ km/s}$$

When **M >> m**, the equation becomes:

$$v_T = \sqrt{(GM/R)} \text{ km/s}$$

Mini-quiz to check your understanding

1. Why would you want to show the orbit of one object with respect to the other?

   a. The is a convenient point of view

   b. Just as an exercise in mathematics

   c. You only consider orbits with respect to the CM

2. How do you determine the orbital velocity of one object relative to the other?

   a. You add their tangential velocity equations relative to the CM

   b. You make **M >> m**

   c. You add their tangential velocities relative to each otherr

3. How can the Earth appear to orbit the Moon?

   a. It is impossible, because the Earth is too big to orbit the Moon

   b. It depends on whether it is a full or new Moon

   c. It happens if you view the Earth from the surface of the Moon

Answers

1a, 2a, 3c

12.3 Direction Convention for Gravitational Motion

Just as there is a convention as to which direction is positive and which is negative for gravity equations, it is also necessary to establish a similar convention for gravitational equations.

The direction of the gravitational attraction of the smaller object toward the larger is defined as positive, while the direction away from the object is negative. Also, the directions are with respect to the starting point of the smaller object.

Since gravitation also allows orbiting around the larger object, a form of radial coordinate system or **r-t** system is used with the zero-point at center of the smaller object.

Vectors and scalars used are similar to those used in the gravity convention.

Gravity coordinate system

The gravity convention for direction considers the motion toward the Earth—in the direction of gravity—as positive and upward or away from Earth as negative. The directions are with respect to the starting point of an object held above the ground.

When studying gravity motion equations, a modified **x-y** coordinate system is used.

We defined a direction convention that set the starting point of an object as the zero-point and designated that downward—toward the ground—was the positive direction and upward was negative.

This meant essentially inverting or flipping the **y**-axis to change the common directions. Motion in the **x**-direction remained the same.

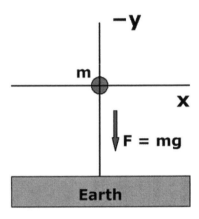

Gravity coordinate system

(See chapter *1.3 Convention for Direction in Gravity Equations* for more information.)

Gravitation coordinate system

Our considerations of gravitation concern two objects in space and the interaction between them.

Although both objects move with respect to the center of mass (CM) between them, it is often more convenient to consider the motion of a smaller object with respect to a larger one. This reduces the need for two motion equations, such that only a single equation is needed.

Since the gravitational force on the smaller object (**m**) is toward the larger object (**M**), that direction is designated as the positive direction. Thus, when object **m** moves away from object **M**, the velocity is in a negative direction.

Because gravitation also allows orbiting around the larger object, a form of radial-tangential coordinate system—or **r-t** system—is used instead of the **x-y** coordinate system of gravity.

The result is that the coordinate system is inverted, similar to that in gravity.

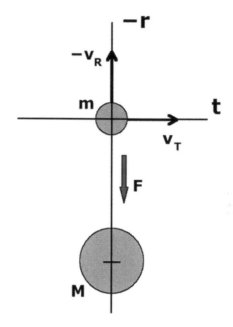

Gravitation coordinate system

Gravitational vectors and scalars

Vectors are divided into those along the axis and those that are perpendicular to the axis of rotation. Scalar quantities are similar to those used in the gravity convention.

Radial vectors

Vectors used in gravitation equations along the axis between the two objects include:

- **F**: Gravitational force
- **R**: Radial displacement
- v_R: Radial velocity

Velocity and displacement vectors pointing away from the CM are negative vectors.

Tangential vectors

Linear tangential velocity (v_T) is the primary tangential vector used for gravitational orbit equations. This is a straight-line velocity that is perpendicular to the axis of rotation and tangent to the curved path of the object.

There is no straight-line displacement tangential vector.

Pseudo-vectors that follow a curved path include:

- ω: Angular velocity
- θ: Angular displacement

Scalars

Scalar quantities related to the vectors in the gravitational equations include:

s: Speed

d: Distance

l: Section of curved path

m: Mass

All scalar magnitudes are positive numbers or quantities.

Magnitude of vectors

The magnitude or absolute value of a vector is a scalar quantity. For example, the absolute value of the velocity vector is the speed of the object:

$$|{-v_R}| = s$$

where

- $-v_R$ is the radial velocity away from the larger object
- $|{-v_R}|$ is the absolute or positive value of the velocity, independent of the direction = speed

Summary

The gravitational direction convention is similar to the convention used for gravity equations. The direction of the gravitational attraction of the smaller object toward the larger is defined as positive, while the direction away from the object is negative. Directions are with respect to the starting point of the smaller object.

A form of radial coordinate system or **r-t** system is used with the zero-point at center of the smaller object. Vectors and scalars used are similar to those used in the gravity convention.

Mini-quiz to check your understanding

1. Why are the direction conventions for gravity and gravitation equations similar?

   a. It is a strange coincidence

   b. Gravity is an approximation of gravitation

   c. They are only similar when the masses of the objects are the same

2. Why is the larger object used as a frame-of-reference?

   a. Because the larger object has no center of mass

   b. Smaller objects move faster than larger objects

   c. You usually consider small objects orbiting larger objects

3. When is a velocity vector negative?

   a. When the velocity is moving toward the larger object

   b. It is impossible for velocity to be negative

   c. When the velocity is moving away from the larger object

Answers

1b, 2c, 3c

12.4 Circular Planetary Orbits

Although the gravitational orbits of the planets around the Sun are slightly elliptical, they are typically considered circular for ease of calculations. This is also true for the orbits of moons around the planets, as well as the various satellites in orbit around the Earth.

If you view the orbits of two objects in space with respect to the center of mass (CM) or barycenter between them, they would each seem to be orbiting that CM.

However, typically you view the orbit of the smaller object with respect to the larger object. This reduces two equations into one velocity equation.

Knowing the masses and separation, you can calculate the tangential velocity of one object with respect to the other. When one object is much larger than the other, the orbital equation is simplified.

Examples of calculations are the velocity of the Moon with respect to the Earth, the Earth in orbit around the Sun and the planet Jupiter around the Sun.

> (See chapter *12.1 Derivation of Circular Orbits Around Center of Mass* for more information.)

Moon orbits the Earth

The Moon and Earth are in orbit around the center of mass (CM) between them. That CM is within the surface of the Earth, such that the Earth appears to wobble about the center when viewed from outer space.

Instead of viewing the orbits from outer space, you can view the orbit of the Moon from the Earth. You simply calculate the orbital velocity of the Moon with respect to the Earth.

(See chapter *12.2 Orbital Motion Relative to Other Object* for more information.)

Location of CM

The location of the CM with respect to the Earth is determined from the equation:

$$R_M = mR/(M + m)$$

where

- R_M is the separation between the center of the Earth and the CM in km
- m is the mass of the Moon = $7.348*10^{22}$ kg = $0.073*10^{24}$ kg
- M is the mass of the Earth = $5.974*10^{24}$ kg
- R is the fixed separation between the Earth and the Moon = $3.844*10^5$ km

(See chapter *11.1 Overview of Gravitation and Center of Mass* for more information.)

The calculated value is:

$$R_M = 7.348*10^{22}*3.844*10^5/(5.974*10^{24} + 0.073*10^{24}) \text{ km}$$

$$R_M = 28.246*10^{27}/6.047*10^{24} \text{ km}$$

$$R_M = 4.671*10^3 = 4671 \text{ km}$$

Since the radius of the Earth is about 6371 km, the CM is almost 3/4 of the radius. This results in the Earth appearing to wobble about the CM, as seen from the Moon

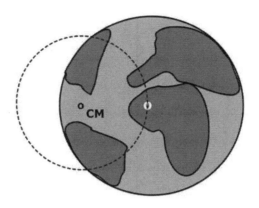

Earth orbits CM between it and the Moon

View from Earth

The Moon's tangential velocity with respect to the Earth is:

$$v_T = \sqrt{[G(M + m)/R]}$$

where

- v_T is the linear tangential velocity of the Moon with respect to the Earth in km/s
- G is the Universal Gravitational Constant = $6.674*10^{-20}$ km^3/kg-s^2

Velocity for circular orbit

The calculation of the velocity of the Moon with respect to the Earth is:

$$v = \sqrt{[G(M + m)/R]} \text{ km/s}$$

$$v = \sqrt{[(6.674*10^{-20})(6.047*10^{24})/(3.844*10^{5})]} \text{ km/s}$$

$$v = \sqrt{(1.050)} \text{ km/s}$$

$$v = 1.025 \text{ km/s}$$

Compare this result from the commonly seen approximation:

$$v_a = \sqrt{(GM/R)}$$

$$v_a = \sqrt{[(6.674*10^{-20})(5.974*10^{24})/3.844*10^{5})]} \text{ km/s}$$

$$v_a = 1.018 \text{ km/s}$$

Earth orbits the Sun

The Earth orbits the Sun in an elliptical orbit that is close to being circular. Since the mass of the Sun is so much greater than the mass of the Earth, the CM between them is almost at the geometric center of the Sun. This simplifies the orbital velocity equation.

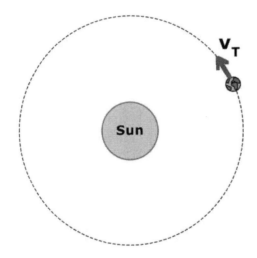

Earth's tangential velocity while orbiting the Sun

Sum of masses

To find the location of the CM, first consider the sum of the masses of the Sun and Earth:

M + m ≈ M

where

- **M** is the mass of the Sun = $1.989*10^{30}$ kg
- **m** is the mass of the Earth = $5.974*10^{24}$ kg = $0.000005974*10^{30}$ kg
- ≈ means "approximately equal to"

$$M + m = 1.989*10^{30} \text{ kg} + 0.000005974*10^{30} \text{ kg}$$

$$M + m \approx 1.989*10^{30} \text{ kg}$$

The sum of the masses is approximately the mass of the Sun.

Location of CM

The equation for the separation between the center of the Earth and the CM is:

$$R_m = MR/(M + m)$$

where

- R_m is the separation between the center of the Earth and the CM
- R is the separation between the centers of the Earth and Sun = $1.496*10^8$ km.

Since $M + m \approx M$, the equation reduces to:

$$R_m = MR/M$$

$$R_m = R$$

Velocity for circular orbit

The equation for the tangential velocity of the Earth orbiting the Sun reduces to:

$$v_T = \sqrt{(GM/R)} \text{ km/s}$$

$$v_T = \sqrt{[(6.674*10^{-20})(1.989*10^{30})/(1.496*10^8)]} \text{ km/s}$$

$$v_T = \sqrt{(8.873*10^2)} \text{ km/s}$$

The tangential orbital velocity of the Earth with respect to the Sun is:

$$v_T = 29.793 \text{ km/s}$$

This corresponds to listed values of the orbital velocity of the Earth. This velocity can also be used to calculate the length of a year or time it takes for one revolution of the Sun.

> (See chapter *12.5 Length of Year for Planets in Gravitational Orbit* for more information.)

Jupiter orbits the Sun

Jupiter is the largest planet in our Solar System. Although the mass of the Sun is much greater than that of Jupiter, the CM or barycenter between Jupiter and the Sun lies outside the Sun's surface. This means that they would both appear to orbit the CM when viewed from outside the Solar System.

However, the view of from the Earth would be that the Sun is a fixed point and that Jupiter's orbit is with respect to the Sun.

Location of CM

Although the mass of the Sun is much greater than that of Jupiter, the CM or barycenter between Jupiter and the Sun lies outside the Sun's surface. The separation between the Sun's center and the CM is:

$$R_M = mR/(M + m)$$

where

- R_M is the separation between the Sun's center and the barycenter
- R is the separation between the centers of the Sun and Jupiter = $7.785*10^8$ km
- M is the mass of the Sun = $1.989*10^{30}$ kg
- m is the mass of Jupiter = $1.899*10^{27}$ kg

Determine $M + m$:

$$M + m = 1.989*10^{30} \text{ kg} + 1.899*10^{27} \text{ kg}$$

Set exponents to be the same:

$$M + m = 1.989*1030 + 0.001899*10^{30} \text{ kg}$$

M + m = 1.991*10^{30} kg

Calculate **R$_M$**:

R$_M$ = (1.899*10^{27} kg)(7.785*10^8 km)/(1.991*10^{30} kg)

R$_M$ = 7.453*10^5 km

Since the radius of the Sun is 695,500 km = 6.955*10^5 km, the CM is outside the surface of the Sun..

Velocity for circular orbit

The calculation of the tangential velocity of Jupiter with respect to the Sun is:

$\mathbf{v_T} = \sqrt{[G(M + m)/R]}$

$\mathbf{v_T} = \sqrt{[(6.674*10^{-20}*1.991*10^{30}/7.785*10^8)]}$ km/s

$\mathbf{v_T} = \sqrt{(1.707*10^2)}$ km/s

$\mathbf{v_T}$ = 13.065 km/s

This corresponds to the measured mean orbital velocity of Jupiter.

Velocity of Jupiter around the CM

The velocity of Jupiter around the CM between Jupiter and the Sun is calculated from the equation:

$\mathbf{v_{Tm}} = \sqrt{[GM^2/R(M + m)]}$ km/s

$\mathbf{v_{Tm}} = \sqrt{[(6.674*10^{-20}*\{1.989*10^{30}\}^2/}$

$7.785*10^8*1.991*10^{30})]$ km/s

$\mathbf{v_{Tm}} = \sqrt{(26.403*10^{40}/15.500*10^{38})}$ km/s

$\mathbf{v_{Tm}} = \sqrt{(170.342)}$ km/s

$\mathbf{v_{Tm}}$ = 13.052 km/s

This value is slightly different that what is observed.

Summary

The gravitational orbits of the planets around the Sun can be considered circular for ease of calculations. This is also true for the orbits of moons around the planets, as well as the various satellites in orbit around the Earth.

Although, each object is orbiting the center of mass (CM) or barycenter between them, by knowing their masses and separation, you can calculate the tangential velocity of one object with respect to the other. This includes the velocity of the Moon with respect to the Earth, the Earth in orbit around the Sun and the planet Jupiter around the Sun.

Mini-quiz to check your understanding

1. How would the orbital velocity of the Moon be affected if it were further from the Earth?

   a. The velocity would increase

   b. The velocity would be slower

   c. The velocity would stay the same

2. Why is the simple equation for orbital velocity used in the Earth-Sun case?

   a. The Earth is too far from the Sun to use anything else

   b. The simple velocity equation is used to get a more accurate result

   c. The Sun is so much larger than the Earth that the CM is almost at its center

3. Why is the CM between Jupiter and the Sun outside the Sun's surface?

   a. Jupiter has such strong gravity that it pulls the CM toward it

   b. The mass of Jupiter is large enough that the CM is outside the Sun

   c. It is only an illusion that the CM is outside the Sun's surface

Answers

1b, 2c, 3b

12.5 Length of Year for Planets in Gravitational Orbit

You can derive the equation for the time it takes an object in a gravitational orbit to make one revolution, provided you know its orbital velocity, the separation between it and the center of the other object and the mass of each object.

Since the time it takes the Earth to make one revolution around the Sun is called a year, it is convenient to state the orbital period in terms of Earth years or even Earth days.

To simplify the calculations, the equation for the orbital period assumes that the orbit is circular.

The equation can be verified by considering the period of the Moon around the Earth and how long it takes the Earth, as well as the planet Jupiter, to go around the Sun.

Velocity to be in a circular orbit

The velocity of an object in circular orbit around another object is a function of the mass of each object and the separation between their centers.

> (See chapter *12.1 Derivation of Circular Orbits Around Center of Mass* for more information.)

Astronomical objects actually orbit the center of mass (CM) between them. The orbital velocity of one object with respect to the other is a sum of their velocities around their CM.

> (See chapter *12.2 Orbital Motion Relative to Other Object* for more information.)

The equation for the velocity of a circular orbit of one object around another is:

$$v_T = \sqrt{[G(M + m)/R]}$$

where

- v_T is the tangential velocity of the object in orbit in kilometers/second (km/s)
- **G** is the universal gravitational constant = $6.674*10^{-20}$ km³/kg-s²
- **M** and **m** are the masses of the objects in kg
- **R** is the separation in km between the objects, as measured from their centers

Note: Because we are stating velocity in km/s, we converted the units of **G** from N-m²/kg² to km³/kg-s². Also, we consider **R** in km instead of meters

In the case where one object has a much greater mass than the other (**M >> m**), the contribution of **m** is negligible. The velocity equation reduces to:

$$v_T = \sqrt{(GM/R)}$$

Deriving equation for one revolution

Knowing the required velocity to be in orbit allows you to determine the time it takes to make one revolution around the other object. The distance traveled in one revolution is the circumference of the circle of radius **R**:

$$C = 2\pi R$$

where

- **C** is the circumference in km
- π stands for pi and equals 3.142...

But also distance equals velocity times time:

$$C = v_T T$$

where **T** is the time in seconds (s) it takes the object to make one revolution around the larger object; it is also called the orbital period.

Combine the two equations for **C** and then solve for **T**:

$$v_T T = 2\pi R$$

$$T = 2\pi R/v_T \text{ seconds}$$

Substitute $v_T = \sqrt{[G(M + m)/R]}$ in the equation for **T**:

$$T = 2\pi R/\sqrt{[G(M + m)/R]}$$

Multiply by $\sqrt{(R)}/\sqrt{(R)}$

$$T = 2\pi R\sqrt{(R)}/\sqrt{[G(M + m)]}$$

$$T = 2\pi\sqrt{(R^3)}/\sqrt{[G(M + m)]}$$

Thus, the equation for the orbital period is:

$$T = 2\pi\sqrt{[R^3/G(M + m)]} \text{ seconds}$$

Verify units

It is a good practice to verify the units to make sure the equation is correct:

$$T \text{ s} = 2\pi\sqrt{[R^3 \text{ km}^3/(G \text{ km}^3/\text{kg-s}^2)(M \text{ kg} + m \text{ kg})]}$$

$$\text{s} = \sqrt{(\text{km}^3/(\text{km}^3/\text{kg-s}^2)\text{kg})}$$

$$\text{s} = \sqrt{(\text{kg-s}^2/\text{kg})}$$

$$\text{s} = \text{s}$$

Convert from seconds

You usually calculate the orbit in Earth days or years, so you need to convert seconds to a different unit of measurement.

1 minute = 60 seconds

1 hour = 60 minutes

1 day = 24 hours

1 year = 365 days

Equation for number of days

Thus, 1 day = (24 hours)*(60 minutes)*(60 seconds) = $8.640*10^4$ seconds. Divide by the number of seconds per day to get the orbit equation for days:

$$D = 2\pi\sqrt{(R^3/GM)}/8.640*10^4 \text{ days}$$

Since $2\pi = 6.284$, you get:

$$D = 7.273*10^{-5}\sqrt{(R^3/GM)} \text{ days}$$

Equation for number of years

Also, 1 year = (365 days)*(86400 seconds) = 31,536,000 seconds. That can be simplified to $3.154*10^7$ s/yr.

$$Y = [2\pi\sqrt{(R^3/GM)}]/3.154*10^7 \text{ years}$$

$$Y = 1.992*10^{-7}\sqrt{(R^3/GM)} \text{ Earth years}$$

Examples

You can calculate the number of days it takes the Moon to rotate around the Earth and the number of years it takes the Earth and the planet Jupiter to go around the Sun.

> **Note**: Since the separation **R** and mass **M** are not exact, as well as the fact that the orbits are not exactly circular, but are slightly elliptical, the time to complete an orbit is not exact. However, the calculations do come out fairly close to what is experienced

Moon orbits the Earth

The number of days that it takes the Moon to complete one revolution around the Earth is:

$$D = (7.273*10^{-5})\sqrt{[R^3/G(M + m)]} \text{ Earth days}$$

where

- **R** = 3.844*10^5 km (separation between centers)
- **G** = 6.674*10^{-20} km³/kg-s²
- **M** = 5.974*10^{24} kg (mass of Earth)
- **m** = 7.348*10^{22} kg = 0.073*10^{24} kg (mass of Moon)

Thus:

$$\mathbf{R^3} = 5.680*10^{16} \text{ km}^3$$

and

$$\mathbf{M + m} = 5.974*10^{24} \text{ kg} + 0.073*10^{24} \text{ kg}$$

$$\mathbf{M + m} = 6.047*10^{24} \text{ kg}$$

Entering in values:

$$\mathbf{D} = (7.273*10^{-5})\sqrt{[5.680*10^{16}/(6.674*10^{-20}*6.047*10^{24})]}$$

$$\mathbf{D} = (7.273*10^{-5})\sqrt{(14.07*10^{10})}$$

$$\mathbf{D} = 7.273*10^{-5}*3.752*10^{5}$$

$$\mathbf{D} = 27.28 \text{ days}$$

That is close to the average of 27.322 days for the Moon to orbit the Earth, as determined by NASA.

Earth orbits the Sun

You can verify the number of days it takes the Earth to orbit the Sun. The center of mass or barycenter between the Earth and the Sun is almost at the Sun's geometric center.

(See chapter *12.4 Circular Planetary Orbits* for more information.)

Thus, the simple equation for the number of days can be used:

$$\mathbf{D_s} = 7.273*10^{-5}\sqrt{(\mathbf{R^3/GM})} \text{ Earth days}$$

where

- **R** = $1.496*10^8$ km (separation between centers)
- **G** = $6.674*10^{-20}$ km³/kg-s²
- **M** = $1.989*10^{30}$ kg (mass of Sun)

Thus:

R³ = 3.348*10²⁴ km³

and:

$$\mathbf{D}_s = 7.273*10^{-5}\sqrt{[3.348*10^{24}/(6.674*10^{-20}*1.989*10^{30})]}$$

$$\mathbf{D}_s = 7.273*10^{-5}\sqrt{[3.348*10^{24}/(6.674*10^{-20}*1.989*10^{30})]}$$

$$\mathbf{D}_s = 7.273*10^{-5}\sqrt{(25.221*10^{12})}$$

$$\mathbf{D}_s = 7.273*10^{-5}*5.022*10^6$$

$$\mathbf{D}_s = 365.25 \text{ days}$$

This value corresponds to most measurements.

The number of years for the Earth to orbit the Sun is:

$$\mathbf{Y} = 1.991*10^{-7}\sqrt{(\mathbf{R^3/GM})}$$

$$\mathbf{Y} = (1.991*10^{-7})(5.022*10^6) \text{ years}$$

$$\mathbf{Y} = 0.999 \text{ years}$$

It takes one year for the Earth to orbit the Sun.

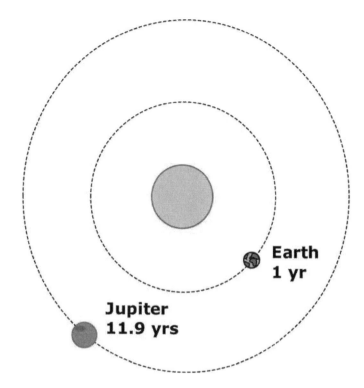

Earth and Jupiter orbit the Sun

Jupiter orbits the Sun

The average separation between the planet Jupiter and the Sun is $R = 7.786*10^{11}$ m. Thus:

$R^3 = 472*10^{33}$ m^3

The mass of the Sun is $M = 1.989*10^{30}$ kg. Thus:

$GM = (6.67*10^{-11})(1.9891*10^{30}) = 13.27*10^{19}$ m^3/s^2

$R^3/GM = 472*10^{33}/13.27*10^{19} = 35.57*10^{14}$ s^2

$\sqrt{(R^3/GM)} = 5.96*10^7$ s

Substitute into $Y = 2*10^{-7}\sqrt{(R^3/GM)}$:

$Y = (2*10^{-7})(5.96*10^7)$ years

Y = 11.9 years

This corresponds to the measured time it takes Jupiter to orbit the Sun.

Summary

If you know the velocity of an object in orbit and its separation from the center of the much larger object, as well as the mass of the larger object, you can calculate how long it takes to make one revolution in Earth days or Earth years.

The equation can be verified by considering the period of the Moon around the Earth and how long it takes the Earth and Jupiter to go around the Sun.

Mini-quiz to check your understanding

1. If many orbits are elliptical, why is the circular orbit equation used?

   a. The velocity for an ellipse is 2/3 the velocity of a circle

   b. Values are easier to calculate and are close enough for most purposes

   c. No one is really sure why scientists continue to use the circular equation

2. Why is the orbital period converted into Earth years?

   a. It is because orbits are not true circles

   b. It depends on what planet you are talking about

   c. It is a convenient way to relate to the time it takes to orbit

3. What would the Earth's orbital period be if the it was 4 times as far away than it is now?

   a. The orbital period would be the same

   b. The orbital period would be 8 years (2*2*2)

   c. The orbital period would be 64 years (4*4*4)

Answers

1b, 2c, 3b

12.6 Effect of Velocity on Orbital Motion

Although orbital motion of two objects in space can be with respect to the center of mass (CM) between them, it is often more convenient to consider the orbital motion of the smaller object with respect to the larger of the two.

A circular orbit requires a specific tangential velocity at the given separation, with no initial radial velocity to affect the orbit. This situation can occur when two celestial bodies pass near each other at the correct velocities and separation, thus resulting in one orbiting the other.

Of more immediate interest is the case of sending up a rocket and aiming it so that its tangential velocity results in a circular orbit.

If the tangential velocity is less than for a circular orbit, the object may go into a small elliptical orbit or even fall into the larger object.

If the tangential velocity is greater than for a circular orbit, the object may go into a large elliptical orbit or fly off into space in a parabolic or hyperbolic path.

The equations for a circular orbit and a parabolic path are the starting points in considering the effect of velocity. Also, if the mass of one object is much greater than the other, the equations can be simplified.

Starting points for orbital equations

The path the object follows depends on its tangential velocity. The velocity equations for a circular orbit and for a parabolic path are starting points for examining the relationship between velocity and orbital motion.

Circular orbit

The tangential velocity equation for a circular orbitwith respect to the other object is:

$$v_T = \sqrt{[G(M + m)/R]}$$

where

- v_T is the tangential velocity in km/s
- **M** and **m** are the masses of the two objects in kg
- **G** is the Universal Gravitational Constant = $6.674*10^{-20}$ km^3/kg-s^2
- **R** is the total separation between the centers of the two objects in km

If one mass is much greater than the other (**M >> m**), the equation reduces to:

$$v_T = \sqrt{(GM/R)}$$

Parabolic path

The equation for a parabolic path is:

$$v_T = \sqrt{[2G(M + m)/R]}$$

Likewise, if **M >> m**, the equation becomes:

$$v_T = \sqrt{(2GM/R)}$$

Orbital paths of object

The possible orbital paths of an object are:

- Small ellipse
- Circular orbit
- Large ellipse

Small elliptical path

When the tangential velocity of an object is less than required for a circular orbit, the object will follow a small elliptical path.

$$0 < v_T < \sqrt{[G(M + m)/R]}$$

If $v_T = 0$, only the gravitational force along the radial axis would affect the falling object, such that the objects would collide. If v_T was slightly greater than zero and both objects were point-masses, the falling object would travel in an elliptical orbit, just missing the other object.

However, if the objects have substantial size, it is possible that the orbiting object will collide with the fixed object. An example of this can be seen with Newton's Cannon, where the cannonball is not propelled fast enough to go into orbit.

(See chapter *6.6 Gravity and Newton's Cannon* for more information.)

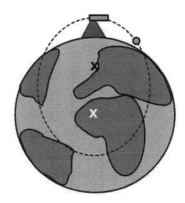

Cannonball does not have enough velocity to go into orbit

Also, since the mass of the cannonball is so small compared to the mass of the Earth, the approximation can be made:

$$0 < v_T < \sqrt{(GM/R)}$$

Circular orbit

At a specific tangential velocity for the given masses and separation, one object will go in a circular orbit with respect to the other object according to the equation:

$$v_T = \sqrt{[G(M + m)/R]}$$

As noted before, we assume there is no initial radial velocity that would complicate things and affect the shape of the orbit.

Large elliptical orbit

If the velocity of the object is greater than required for a circular object but is less than that of a parabolic path, the object will travel in a large elliptical orbit around the other object:

$$\sqrt{[G(M + m)/R]} < v_T < \sqrt{[2G(M + m)/R]}$$

If **M >> m**:

$$\sqrt{(GM/R)} < v_T < \sqrt{(2GM/R)}$$

Go out into space

If the velocity is great enough, the object will escape the gravitational force and fly out into space. It will travel in a parabolic or hyperbolic path, depending on the initial tangential velocity.

Parabolic path

If the tangential velocity is such that:

$$vT = \sqrt{[2G(M + m)/R]}$$

the object will follow a parabolic path and go off into space. This velocity is called the *gravitational escape velocity*.

> (See chapter *13.1 Overview of Gravitational Escape Velocity* for more information.)

Likewise, if **M >> m**:

$$v_T = \sqrt{(2GM/R)}$$

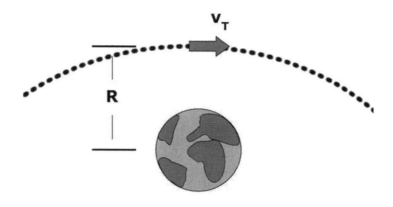

Rocket follows parabolic path

Hyperbolic path

If the tangential velocity is is greater than required for a parabolic path, the object will follow a hyperbolic path and go off into space:

$$v_T < \sqrt{[2G(M + m)/R]}$$

If $M \gg m$:

$$v_T < \sqrt{(2GM/R)}$$

In this situation, the velocity exceeds the escape velocity.

Summary

The tangential velocity of one object with respect to another determines whether the object collides with the other object, goes into orbit or flies off into space.

The velocity equations for circular orbits and parabolic paths are the starting points for establishing the effect of tangential velocity on orbital motion.

The velocity equations are:

Collide or small elliptical orbit:

$$0 < vT < \sqrt{[G(M + m)/R]}$$

Circular orbit:

$$vT = \sqrt{[G(M + m)/R]}$$

Large elliptical orbit:

$$\sqrt{[G(M + m)/R]} < vT < \sqrt{[2G(M + m)/R]}$$

Parabolic path:

$$vT = \sqrt{[2G(M + m)/R]}$$

Hyperbolic path:

$$vT < \sqrt{[2G(M + m)/R]}$$

If the mass of one object is much greater than the other, the equations are simplified.

Mini-quiz to check your understanding

1. What are factors that determine whether the objects will collide?

   a. Tangential velocity and size of the objects

   b. Objects never collide when moving in space

   c. Separation of objects

2. What happens when the tangential velocity is slightly greater than required for a circular orbit?

   a. The object will spiral into the ground

   b. The object will go into an elliptical orbit

   c. The velocity will stabilize so that a circular orbit is maintained

3. What happens when the velocity is greater than the escape velocity?

   a. The object takes a parabolic path

   b. The object takes a hyperbolic path

   c. It is impossible for a velocity to be greater than the escape velocity

Answers

1a, 2b, 3c

Part 13: Escape Velocity

One application of gravitation and forces in the Universe is the velocity required to escape the gravitation from the Earth or other celestial body

Part 13 chapters

Chapters in Part 13 include:

13.1 Overview of Gravitational Escape Velocity

This chapter provides an overview of gravitational escape velocity, explaining what it is and what the requirements are for reaching the escape velocity. It gives the equation for calculating the escape velocity and provides some examples.

13.2 Gravitational Escape Velocity Derivation

The derivation of the escape velocity from a planet is determined in this chapter, using the potential and kinetic energies of the object. The general escape velocity equation is stated.

13.3 Gravitational Escape Velocity with Saturn V Rocket

This chapter demonstrates how the Saturn V rocket launched the Apollo 11 spacecraft to the Moon. It shows how the rocket was able to approach the escape velocity from the Earth.

13.4 Effect of Sun on Escape Velocity from Earth

The gravitational force of the Sun can affect the escape velocity of a rocket from Earth. This chapter provides the calculations with and without the effect of the Sun.

13.5 Gravitational Escape Velocity for a Black Hole

This chapter shows how a Black Hole has an escape velocity greater than the speed of light. From that fact, the virtual size of the Black Hole can be determined.

13.1 Overview of Gravitational Escape Velocity

Gravitational escape velocity is the release velocity of a freely moving object that has been accelerated away from the surface of a sun, planet or moon, such that this initial velocity is sufficient to prevent it from being overcome by gravitational force and falling back to the surface.

Although the escape velocity can be considered with respect to the center of mass (CM) between the objects, it is usually measured with respect to the larger of the two objects.

Also, the mass of the escaping object is considered as much less than the mass of the attracting object.

A simple equation provides the escape velocity as a function of the initial separation of the objects and the mass of the larger body. There are several conditions for the equation to be valid.

The equation allows you to calculate the escape velocity from any celestial body, provided you know the body's mass and radius, as well as the altitude of the object. Applications of the equation include the calculations of the escape velocity from the Earth, Moon and Sun.

Escape velocity equation

When an object is projected at a sufficient velocity in a direction away from a much larger object at some altitude, it can escape the gravitational attraction between the two and fly off into space.

This initial velocity is called the escape velocity.

The direction of the velocity can be in the radial or vertical direction, in the tangential direction or directions in between, as long as it results in a direction away from the larger object.

Considering our convention that velocity vectors moving in the opposite direction of gravitation are negative, the standard gravitational escape velocity equation is:

$$v_e = -\sqrt{(2GM/R)}$$

where

- v_e is the escape velocity in kilometers/second (km/s)
- G is the Universal Gravitational Constant = $6.674*10^{-20}$ km^3/kg-s^2
- M is the mass of the planet or sun in kilograms (kg)
- R is the separation in km between the objects at the point of release

Escape speed

Since the equation applies for a variety of directions resulting in motion away from the larger body, some sources refer to it as escape speed and use the magnitude of the velocity equation:

$$s_e = \sqrt{(2GM/R)}$$

where s_e is the escape speed.

However, this implies that direction is not a factor. It overlooks the fact that the projected object may crash into the larger body when sent in some directions.

> **Note**: I feel the velocity equation is more correct, and the negative sign indicates the object is moving in a direction opposite the gravitational force.

Considering altitude

When considering the separation of the objects, it is often more convenient to use the radius of the larger object plus the altitude or height instead of separation. For example, let:

R = r + h

where

- **r** is the radius of the planet or sun and the center of the object in km
- **h** is the separation from the center of the escaping object to the surface of the planet or sun in km

Thus, the equation becomes:

$$v_e = -\sqrt{[2GM/(r + h)]}$$

This is shown in the illustration below:

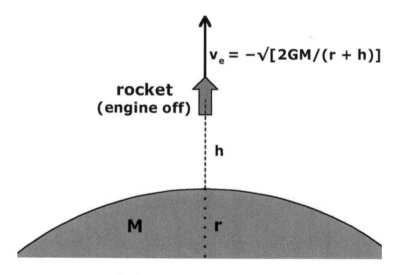

Rocket reaches escape velocity

Mass of object not a factor

Surprisingly, the mass of the object projected upward is not a factor in the escape velocity. But the escape velocity does depend on the mass of the body from which it is escaping.

Conditions and assumptions

There are several other conditions or assumptions for the gravitational escape velocity equation:

Freely moving

High altitude necessary for escape

Effect of planetary rotation ignored

Effect of other objects not included

Freely moving

Although the object or rocket may be accelerated up to the escape velocity, any means of propulsion is turned off and the object is moving freely.

High altitude required to reach velocity

It is usually necessary for an object to reach a high altitude before it achieves the gravitational escape velocity of the celestial body.

Acceleration from surface

If an object is accelerated from the surface of the planet or sun, the object will often need to travel a great distance to some high altitude in order to reach a sufficient velocity to escape.

> **Note**: Some textbooks refer to *surface escape velocity*. Unfortunately, that concept is incorrect, because objects of any size do not instantaneously accelerate to such a high velocity. Also, many calculations use the escape velocity from gravity, which also are measured from the surface.
>
> (See chapter *6.7 Escape Velocity from Gravity* for more information.)

Higher altitudes reduces required velocity

The further the object is from the center of the celestial body, the lower the required escape velocity.

In the situation of a rocket blasting off from the Earth, Moon or some planet, the point where the engines shut off and the rocket starts coasting is where the escape velocity is determined. The higher the rocket goes before the engines shut off, the lower the required initial velocity to escape.

Overcoming air resistance

Also, when escaping the Earth's gravitation, air resistance must be overcome. Often a rocket will first go into orbit at some high altitude—typically, around 190 km or 120 mi, where air resistance is no longer a factor. Then it will blast off into space to reach the escape velocity.

> (See chapter *13.3 Gravitational Escape Velocity with Saturn V Rocket* for more information.)

Ions and subatomic particles escaping from Sun's gravitation are usually sent upward in turbulent solar storms until they reach the escape velocity for the altitudes reached.

Rotation not included

The effect of planet rotation and orbital motion are not figured into the escape velocity equation we are using. Those factors can decrease or increase the escape velocity but also complicates calculations.

For example, the escape velocity of an object in the direction of the rotation of the Earth is less than when not calculating the rotation. It also varies with the latitude on Earth from which a rocket is fired. That is why it is preferred to launch rockets near the Earth's equator.

Effect of other objects not included

The effect of gravitation forces from other objects is not considered. Gravitation from the Sun on an object leaving the Earth influences the escape velocity but is not included in our simple equation.

> (See chapter *13.4 Effect of Sun on Escape Velocity from Earth* for more information.)

Common escape velocities

You can use the escape velocity equation to determine the necessary velocity for an object projected upward to escape the Earth, Moon or Sun.

Escape velocity from Earth

The Saturn V rocket that was used to go to the Moon employed its first two stages to reach a speed that was close to the escape velocity at an altitude of about 191 km (119 mi).

What is the required escape velocity at that altitude, such that the rocket will continue rising without more propulsion and not fall back to Earth?

Solution

The radius of the Earth is about **r** = 6371 km or $6.371*10^3$ km, and its mass is approximately **M** = $5.974*10^{24}$ kg.

Add the radius of the Earth to the altitude of the rocket:

R = (6371 + 191) = 6562 km

Substitute the values into the escape velocity equation:

$\mathbf{v_e} = -\sqrt{(2GM/R)}$

$\mathbf{v_e} = -\sqrt{[2*(6.674*10^{-20}\ km^3/kg\text{-}s^2)*(5.974*10^{24}\ kg)/(6.562*10^3\ km)]}$

$\mathbf{v_e} = -\sqrt{(121.519\ km^2/s^2)}$

$\mathbf{v_e} = -11.02$ km/s

In other words, the calculated escape velocity from 191 km above the Earth's surface is about 39,685 km/hr or 24,684 mi/hr.

This velocity is less than the *surface escape velocity* from gravity of 11.2 km/s, which is also an impossible scenario.

Escape velocity from the Moon

Suppose a rocket landed on the Moon and then blasted off to return to Earth. The rocket's engines only operated for about 10 km to build up its speed. What velocity must the rocket attain to escape the Moon?

Solution

The approximate radius of the Moon is 1737 km and its mass is about $7.3*10^{22}$ kg. The separation between the center of the Moon and the center of the rocket is:

$$\mathbf{R} = (1737 + 10) = 1.747*10^3 \text{ km}$$

Substitute values into the escape velocity equation:

$$\mathbf{v_e} = -\sqrt{[2*(6.674*10^{-20})*(7.347*10^{22})/(1.747*10^3)]} \text{ km/s}$$

$$\mathbf{v_e} = -\sqrt{(5.613)} \text{ km/s}$$

$$\mathbf{v_e} = -2.37 \text{ km/s}$$

Thus, the escape velocity from near the surface of the Moon is about 8529 km/hr or 5305 mi/hr.

Escape velocity from the Sun

Solar flares usually reach about 5000 km from the Sun's surface, as seen from telescopes on Earth. These flares project subatomic particles out into space as a coronal mass ejection or massive burst of solar wind.

If the storm pushed particles 100,000 km (10^5 km) until they reached their escape velocity from the Sun, what would that velocity be?

Solution

The radius of the Sun is about $6.955*10^5$ km and its mass is approximately $1.989*10^{30}$ kg. The separation between the center of the Sun and a subatomic particle is:

$$R = (6.955 * 10^5 + 1 * 10^5) = 7.955 * 10^5 \text{ km}$$

Substitute values into the escape velocity equation:

$$v_e = -\sqrt{[2 * (6.674 * 10^{-20}) * (1.989 * 10^{30})/(7.955 * 10^5)]} \text{ km/s}$$

$$v_e = -\sqrt{(33.374 * 10^4)} \text{ km/s}$$

$$v_e = -5.777 * 10^2 = 577.7 \text{ km/s}$$

Thus, the escape velocity at 100,000 km from the surface of the Sun is 2,079,720 km/hr or 1,293,586 mi/hr.

Summary

Gravitational escape velocity is the velocity of an object that is sufficient to escape the gravitation of a much larger body, so that it flies off into space.

The escape velocity equation is:

$$v_e = -\sqrt{(2GM/R)}$$

or

$$v_e = -\sqrt{[2GM/(r + h)]}$$

The escape velocity equation assumes that the object is not being propelled, it is released at a high altitude and rotation of the large body is not considered. Also, gravitation from other objects is not considered.

The equation allows you to calculate the escape velocity from any planet, moon or sun. Applications of the equation include the calculations of the escape velocity from the Earth, Moon and Sun..

Mini-quiz to check your understanding

1. Why is determining the escape velocity from the surface of the Earth invalid?

   a. Gravity is much greater at the surface of the Earth

   b. The object cannot instantaneously reach the escape velocity

   c. It is valid if you set **R** = 0

2. How does the escape velocity change if you double the mass of the rocket?

   a. The escape velocity doubles

   b. The rocket is then unable to escape

   c. Escape velocity is independent of the mass of the rocket

3. If a star had twice the mass of the Sun and twice the radius, how would its escape velocity compare?

   a. The star's escape velocity would be twice as large

   b. The escape velocities would be the same

   c. The star's escape velocity would be half as large

Answers

1b, 2c, 3b

13.2 Gravitational Escape Velocity Derivation

In order to overcome the gravitational pull of a planet, moon or sun, an object must be accelerated to the gravitational escape velocity for that celestial body at a given altitude above the body's surface. It is then assumed to be moving freely with only the gravitational force from the larger object being applied.

Typically, the projected object has a much smaller mass than the body from which it is escaping, such that the center of mass between them is located close to the geometric center of the larger body. Thus, instead of both objects moving away from the center of mass, the smaller object is considered to be moving away from the larger body.

The equation for the escape velocity can be derived by applying the *Law of Conservation of Energy*. This Law states that the total of the object's potential and kinetic energy is a constant. The escape velocity equation is also a function of the separation between the centers of the object and the celestial body from which it is escaping.

The derivation starts with the initial *gravitational potential energy* at the given altitude and the initial *kinetic energy* of the object. This total initial energy is then compared with the sum of the potential and kinetic energies at an infinite separation, in order to determine the escape velocity equation.

Initial energy of the object

Once an object—such as a rocket—has reached a sufficient velocity above the surface of a moon, planet or sun and is no longer being powered, it has an *initial gravitational potential energy* and an *initial kinetic energy*.

Initial potential energy

The gravitational potential energy between two objects at some separation is defined as the work required to move them from a *zero reference point* to that given separation. The zero reference point from gravitation is at an infinite separation.

However, since there is a great difference in mass between the objects, the potential energy can also be considered as the work required to move the smaller object to the zero reference point. The separation will still be infinite.

> **Note**: You can also consider the motion of the smaller object with respect to the larger object instead of with respect to the center of mass. In either case, since **M >> m**, the results are the same.

The initial gravitational potential energy between two objects at some separation is:

$$PE_i = -GMm/R_i$$

where

- PE_i is initial gravitational potential energy in kg-km^2/s^2
- G is the Universal Gravitational Constant = $6.674*10^{-20}$ km^3/kg-s^2
- M is the mass of the attracting object in kilograms (kg)
- m is the mass of the escaping object in kg
- M is much greater than **m** (**M >> m**)
- R_i is the initial separation or distance between the centers of the objects in kilometers (km)

Note: Since the work required to move an object from the zero reference point of **PE** to some point in space is negative, the **PE** at that point is also considered negative.

(See chapter *9.4 Gravitational Potential Energy* for more information.)

Also note: Since escape velocity is usually stated in km/s, the units of **PE** and **R** have been changed to kilometers. Likewise, the value to **G** has been changed to reflect use of kilometers.

Initial kinetic energy

The initial kinetic energy of an object projected at some velocity away from the Earth or other astronomical body is:

$$KE_i = mv_e^2/2$$

where

- KE_i is the initial kinetic energy in kg-km^2/s^2
- **m** is the mass of the object in kg
- v_e is the initial velocity—and thus the escape velocity—in km/s

Illustration of factors

The following picture shows the relationship of factors involved:

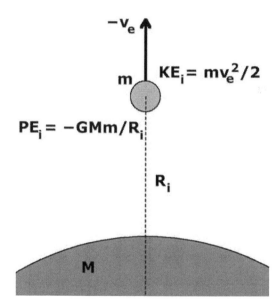

Factors involved in gravitational escape velocity

Note: Our direction convention states that a velocity vector that is "up" or away from the larger mass is in the negative direction.

(See chapter *12.3 Direction Convention for Gravitational Motion* for more information.)

Total initial energy

The total initial energy is the sum of the potential and kinetic energies at the release point:

$$TE_i = KE_i + PE_i$$

$$TE_i = mv_e^2/2 - GMm/R_i$$

Final PE and KE

Gravitational fields hypothetically extend to infinity. Thus, if the initial velocity is great enough, the object will travel to an infinite separation and thus "escape" the gravitational force.

Potential energy at infinity

The object's potential energy at an infinite separation or displacement is:

$$PE_\infty = -GMm/R_\infty$$

where

- PE_∞ is the gravitational potential energy at infinity
- R_∞ is the infinite distance between the centers of the objects

Since $R_\infty = \infty$ (infinity), then $PE_\infty = 0$.

Kinetic energy at infinity

The object's kinetic energy at an infinite displacement is:

$$KE_\infty = mv_\infty^2/2$$

where

- KE_∞ is the final kinetic energy
- v_∞ is the final velocity

At infinity, the velocity of the object is zero: $v_\infty = 0$. Thus $KE_\infty = 0$.

Total final energy

Since the kinetic energy is moving upward and the potential energy is acting downward, the total energy at the initial position is:

$$TE_\infty = KE_\infty + PE_\infty$$

$$TE_\infty = 0 + 0$$

Escape velocity equation

The *Law of Conservation of Energy* states that the total energy of a closed system remains constant. In this case, the closed system consists of the two objects with the gravitational force between them and no outside energy or force affecting either object.

Thus the total final energy—potential energy plus kinetic energy—must equal the total initial energy:

$$TE_i = TE_\infty$$

$$KE_i + PE_i = 0$$

$$mv_e^2/2 - GMm/R_i = 0$$

$$mv_e^2/2 = GMm/R_i$$

$$v_e^2 = 2GM/R_i$$

Taking the square root of each expression results in:

$$v_e = \pm\sqrt{(2GM/R_i)}$$

Considering our gravitational convention for direction, v_e is upward or away from is upward or away from the other object and is thus negative:

$$v_e = -\sqrt{(2GM/R_i)}$$

Note: Although convention-wise, the negative version of the equation is correct, most textbooks will give the positive version of the equation. You should be aware of this fact.

Altitude factor

The equation can also be written considering the altitude of the escaping object:

$$v_e = -\sqrt{[2GM/(r + h)]}$$

where

- **r** is the radius of the celestial body in km
- **h** is the altitude or separation from the surface of the body in km

The altitude factor is necessary since the escaping object must accelerate over some displacement to reach the escape velocity.

Summary

The derivation of the gravitational escape velocity of an object from a much larger mass is achieved by comparing the potential and kinetic energy values at some given point with the values at infinity, applying the Law of Conservation of Energy. The equation for the gravitational escape velocity is:

$$v_e = -\sqrt{(2GM/R_i)}$$

Taking altitude into account, the equation can be written as:

$$v_e = -\sqrt{[2GM/(r + h)]}$$

Mini-quiz to check your understanding

1. At what point is the escape velocity of a rocket determined?

    a. At the altitude that the rocket engine turns off

    b. When the rocket leaves the ground

    c. At the point that **KE = PE**

2. Why must the final total energy be considered at infinity?

    a. No one is really sure why, except that it works

    b. It is to make the escape velocity much larger

    c. Gravitation hypothetically extends to infinity

3. What law is used to determine the escape velocity equation?

    a. The Law of Conservation of Mass

    b. The Law of Conservation of Energy

    c. Einstein's Law of Relatives

Answers

1a, 2c, 3b

13.3 Gravitational Escape Velocity with Saturn V Rocket

Saturn V was an American rocket used in the NASA space programs. The Saturn-powered flight of Apollo 11 to the Moon gives a typical scenario of a rocket reaching near the gravitational escape velocity of the Earth at a high altitude.

In Apollo 11, the Saturn V rocket burned through three stages, each reaching higher velocities and altitudes. From the NASA data, you can see the altitude and velocity necessary for orbiting the Earth, as well as the altitude for the escape velocity.

Using calculations of orbital or tangential velocity and the required escape velocity, you can see that projecting the rocket in the direction of the parking orbit was the most efficient means of reaching the higher velocity.

Escaping Earth's gravitation

It takes a great effort to project an object upward so that it will escape the Earth's gravitation.

The object—usually a rocket—must accelerate from zero to over 32,000 kilometers per hour (km/h) or 20,000 miles per hour (mph).

This requires traveling to high altitudes before the speed is attained. At lower altitudes, air resistance is a major obstacle. It not only holds back the rocket, but it can also cause overheating at high velocities.

The often-quoted surface escape velocity from gravity of 11.2 km/s is not valid, because it assumes instantaneous acceleration. It also ignores the potential destructive nature from air resistance at such a high velocity and low altitude.

Rockets such as Saturn V typically consist of several sub-rockets or stages, each taking the rocket to a higher velocity and altitude in an effective manner. When the first stage expends its fuel, it separates from the rocket, thus reducing the rocket's weight. Then the second stage ignites, pushing the rocket to even greater heights. Finally, the third stage completes the acceleration.

Saturn V stages

The Saturn V rocket that propelled Apollo 11 to the Moon is an example of reaching near the escape velocity by using three stages.

Stage S-IC

When Saturn V blasted off from the Earth, the first stage burned for 2.5 minutes, lifting the rocket to an altitude of 68 km (42 miles) and a speed of 2.76 km/s (9,920 km/h or 6,164 mph). This speed was much less than the escape velocity.

Stage S-II

After the S-IC stage separated from the Saturn V rocket, the S-II second stage burned for 6 minutes. This propelled the rocket to an altitude of 176 km (109 miles) and a speed of 6.995 km/s (25,182 km/h or 15,647 mph). This speed is close to the orbital velocity for that altitude.

Stage S-IVB

After the S-II stage separated from the rocket, the third stage burned for about 2.5 minutes. It then cut off, and the Apollo 11 went into a "parking orbit" at an altitude of 191.2 km (118.8 miles). Its velocity was 7.791 km/s (28,048 km/h or 17,432 mph).

Blasting away from Earth

After several orbits around the Earth, the rocket's engines re-ignited, and it blasted off for what they call *translunar injection*. According to NASA data, Saturn V reached an altitude of

334.436 km and a speed of 10.423 km/s, at which time the engines were shut down.

This velocity was less than the escape velocity for that altitude (see the *Calculation section* below), but it was sufficient to take Apollo 11 to the Moon. The gravitational attraction from the Moon, facilitated its motion.

Seeking escape velocity

If the attraction from the Moon was not a factor and the propose was for Saturn V to reach the escape velocity, the rocket could have blasted off in a vertical direction from the Earth or simply accelerated along the tangential direction or motion.

Going in a vertical direction would require the rocket to accelerate from zero velocity to the escape velocity. However, going up to the escape velocity in the radial direction required much less energy.

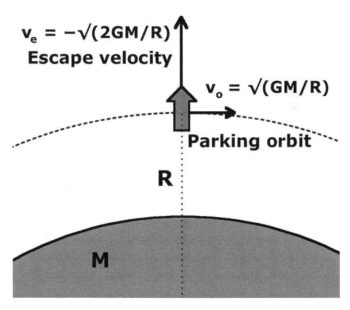

Saturn V parking orbit and required escape velocity

Calculations

Using the data given, you can calculate the orbital and escape velocities.

Parking orbit velocity

At an altitude of 191.2 km, Apollo 11 went into a parking orbit. The stated NASA velocity was 7.791 km/s. Compare this velocity with a calculated orbital velocity:

$$v_T = \sqrt{(GM/R)}$$

where

- v_T is the tangential orbital velocity in km/s
- **G** is the Universal Gravitational Constant = $6.674 * 10^{-20}$ km^3/kg-s^2
- **M** is the mass of the Earth = $5.974 * 10^{24}$ kg
- **R** is the radius of the Earth (6371 km) plus the altitude of the rocket (191.2 km)

(See chapter *12.4 Circular Planetary Orbits* for more information.)

$$R = 6371 \text{ km} + 191.2 \text{ km} = 6562.2 \text{ km}$$

$$v_T = \sqrt{(6.674 * 10^{-20} * 5.974 * 10^{24} / 6562.2)} \text{ km/s}$$

$$v_T = \sqrt{(60.759)} \text{ km/s}$$

$$v_T = 7.795 \text{ km/s}$$

Since the mass and radius of the Earth are approximations, this calculated value for the orbital velocity is sufficiently close to the measured velocity.

Gravitational escape velocity

The escape velocity equation is:

$$v_e = -\sqrt{(2GM/R)}$$

The negative value v_e is the radial escape velocity in km/s. It is negative because the direction is away from the center of mass. Since we are comparing the radial and tangential velocities, it is often more convenient to use s_e, the positive magnitude or escape speed.

(See chapter *12.3 Direction Convention for Gravitational Motion* for more information.)

At an altitude of 334.436 km, Apollo 11 had attained a speed of 10.423 km/s. The calculated gravitational escape velocity or speed at that altitude ($R = 6705.4$ km) is:

$s_e = \sqrt{(2*6.674*10^{-20}*5.974*10^{24}/6705.4)}$ km/s

$s_e = \sqrt{(118.920)}$ km/s

$s_e = 10.905$ km/s

Rocket did not have to reach escape speed

Since the rocket was going to the Moon, its velocity (10.423 km/s) did not have to be the escape velocity (10.905 km/s) for that altitude, especially since the gravitational force of the Moon also had an effect on the rocket.

Note: Several sources state that the rocket reached the escape velocity of 11.2 km/s. However, that is incorrect, since 11.2 km/s is the surface escape velocity, which does not apply to the high altitude.

It is possible that calculations were made without referring to the NASA data and were done simply assuming the escape velocity was 11.2 km/s.

Direction of rocket

If the rocket was propelled in a vertical or radial direction, it would have to accelerate from 0 km/s in that direction to 10.905 km/s.

However, if the rocket went in the tangential direction, it would accelerate from v_T to v_s. The resulting change in velocity would be 10.905 - 7.795 = 3.11 km/s, which would require much less energy.

This shows that accelerating in the direction of the parking orbit was more efficient way to reach the escape velocity.

Summary

The Saturn V rocket was used to power Apollo 11 to the Moon. The flight shows how a rocket reaches close to the gravitational escape velocity of the Earth at a high altitude.

The Saturn V rocket went through three stages, until it reached an altitude of 191.2 km to go into a parking orbit. The rocket then accelerated to an altitude of 334.436 km, where it was close to the calculated escape velocity of 10.905 km/s.

Calculations for the velocity of a circular orbit at that altitude closely correspond to the NASA-measured velocity. The calculations also show that going in the direction of the parking orbit was a more efficient way to reach the escape velocity.

Mini-quiz to check your understanding

1. What would happen to a rocket that attained the escape velocity at a very low altitude?

    a. It would escape from Earth earlier than expected

    b. There would be the danger to people below

    c. The air resistance would overheat the outer shell of the rocket

2. Why were three stages used in Saturn V?

    a. The method effectively reduced its weight as fuel was expended

    b. Three different companies made the stages

    c. No one really knows for sure why they did that

3. Why didn't Apollo 11 reach the gravitational escape velocity?

    a. Apollo 11 would burn up if it ever reached the escape velocity

    b. It wasn't necessary, since Apollo 11 was only going to the Moon

    c. Apollo 11 exceeded the escape velocity, but the fact was never disclosed

Answers

1c, 2a, 3b

13.4 Effect of Sun on Escape Velocity from Earth

The escape velocity equation allows you to calculate the velocity an object—such as a rocket—must attain in order to completely overcome the gravitational pull of the Earth. However, the equation does not take into consideration the effect of the Sun's gravitation on the escape velocity.

When a rocket blasts off from the Earth in a direction away from the Sun, it must not only escape the Earth's gravitation but also the gravitational pull from the Sun.

Surprisingly, the escape velocity from the Sun at the Earth's surface is greater than the escape velocity from the Earth. The two factors must be combined to give the true escape velocity.

Escape velocity from Earth

The escape velocity from Earth can be calculated from the equation:

$$v_E = -\sqrt{(2GM/R_i)}$$

where

- v_E is the escape velocity from Earth in km/s
- G is the Universal Gravitational Constant = $6.674*10^{-20}$ km³/kg-s²
- M is the approximate mass of the Earth = $5.974*10^{24}$ kg
- R_i is the initial separation between the center of the Earth and the center of mass of the rocket

Note: Since v_E is away from the center of mass between the Earth and the rocket, it is negative, according to our direction convention.

If the mean radius of the Earth is 6371 km and the rocket shut its engines at 334.4 km (such as with the Saturn V rocket), the initial separation from the Earth's center would be:

R_i = 6371 km + 334.4 km = 6705.4 km

(See chapter *13.3 Gravitational Escape Velocity with Saturn V Rocket* for more information.)

Substituting values into the equation results in:

$v_E = -\sqrt{[2*(6.674*10^{-20})*(5.974*10^{24})/(6705.4)]}$ km/s

$v_E = -\sqrt{(118.920)}$ km/s

$v_E = -10.905$ km/s

A rocket would have to achieve this velocity before shutting off its engines, if it were to escape from the gravitational pull of the Earth.

Escape velocity from Sun at Earth

Suppose a rocket blasted off the Earth from its far side, away from the Sun. The escape velocity from the Sun would be:

$v_S = -\sqrt{(2GM_S/D)}$

where

- M_S is the approximate mass of the Sun = $1.988*10^{30}$ kg
- D is the approximate separation between the Sun and the rocket on the far side of the Earth = $1.496*10^8$ km

Note: The separation between the center of the Sun and the center of the Earth is approximately $1.496*10^8$ km. The addition of the 6705.4 km would make a negligible contribution.

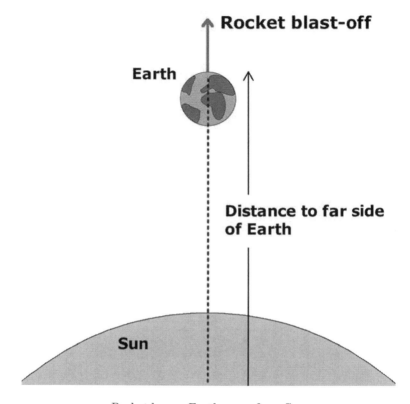

Rocket leaves Earth away from Sun

Substituting values into the equation results in:

$$v_S = -\sqrt{(2*6.674*10^{-20}*1.988*10^{30}/1.496*10^8)} \text{ km/s}$$

$$v_S = -\sqrt{(17.738*10^2)} \text{ km/s}$$

$$v_S = -42.1 \text{ km/s}$$

The escape velocity from the Sun at the Earth's surface is larger than the escape velocity from the Earth itself. This means that the rocket may be able to escape the Earth but would have to go much faster to escape the Sun's gravitation.

Combined escape velocity

If the rocket did not exceeded the Escape velocity from the Earth but not the escape velocity from the Sun, it would move off into space and then soon reverse directions and fall into the Sun.

The combined escape velocity from the Earth, adding in the effect from the Sun, is:

$$v_e = -\sqrt{(v_E^2 + v_S^2)} \text{ km/s}$$

$$v_e = -\sqrt{(10.9^2 + 42.1^2)} \text{ km/s}$$

$$v_e = -\sqrt{(1891.22)} \text{ km/s}$$

$$v_e = -43.488 \text{ km/s}$$

This velocity does not take into account the rotation of the Earth and its orbital velocity, which will affect the escape velocity. This is beyond the scope of our studies. Also, the contribution from the gravitation of the Moon is negligible.

Moving toward the Sun does not apply

The escape velocity concept fails when the rocket blasts from the Earth on the side facing the Sun, because rocket is no longer escaping the Earth's gravitation and moving out to infinity. Instead, the rocket is leaving the Earth and being attracted to the Sun.

Summary

The escape velocity from the Earth does not take into account the escape velocity from the Sun that an object—such as a rocket—must attain in order to completely overcome the gravitational pull of both the Earth and the Sun.

When a rocket blasts off from the Earth in a direction away from the Sun, you must combine the escape velocity from Earth and the escape velocity from the Sun to get the true escape velocity.

Mini-quiz to check your understanding

1. Why must the rocket engines be shut off?

   a. To prevent overheating when achieving escape velocity

   b. The equation is for freely moving objects

   c. To save fuel in case the rocket starts falling

2. Why is the effect of the Sun measured on the far side of Earth?

   a. That is where the Sun's gravitation is pulling against the rocket motion

   b. Rocket blast-offs are always at night

   c. There is no rotation of the Earth in that area

3. What would happen if the rocket did not attain the combined escape velocity?

   a. It really wouldn't matter, since the rocket was moving upward

   b. The rocket would have to reduce its mass

   c. It might escape Earth but then fall into the Sun

Answers

1b, 2a, 3c

13.5 Gravitational Escape Velocity for a Black Hole

A *Black Hole* is a very massive sun or star that has collapsed on itself, such that its gravitational field is so strong than not even light can escape its pull. It is called a "black hole" because that is how it appears to telescopes.

The equation for the gravitational escape velocity of a Black Hole uses the speed of light as the velocity for material trying to escape the star.

The defining separation from the center of a Black Hole is called the *event horizon* or *Schwarzschild radius* and can be determined from the escape velocity equation.

You can also use the equation to calculate the size of the Sun if its mass were compressed enough to make it a Black Hole.

Black Hole escape velocity

The equation for the escape velocity of a Black hole is obtained by substituting the speed of light in the standard escape velocity equation.

Standard escape velocity equation

The standard escape velocity equation of a small object from a celestial body is:

$$v_e = -\sqrt{(2GM/R)}$$

where

- v_e is the escape velocity in kilometers/second (km/s)
- G is the Universal Gravitational Constant = $6.674*10^{-20}$ km^3/kg-s^2

- **M** is the mass of the sun or star in kilograms (kg)
- **R** is the separation between the center of the sun and the center of the object in km

Note: v_e is negative, since the direction is away from the center of the celestial body. Also, since v_e is in km/s, **G** is stated in km^3/kg-s^2 and **R** in km.

Black Hole equation

If the mass of the star was compressed to such a small size or high density that the escape velocity was greater than the speed of light, any particles or objects projected upward from its surface could not escape the gravitational pull:

$$s_e > c$$

where

- s_e is escape speed or magnitude of v_e
- **c** is the speed of light in a vacuum = $2.997*10^5$ km/s

Einstein proved in his *General Theory of Relativity* that light is affected by gravitation. This means that even light or electromagnetic waves could not escape from a Black Hole.

Thus, the escape velocity equation for a Black Hole is:

$$s_e = \sqrt{(2GM/R)} > c$$

This equation means that when the values of **M** and **R** for a celestial object are such that $\sqrt{(2GM/R)}$ is greater than the speed of light, nothing can escape the body—not even light.

Finding the radius for a given mass

An interesting application of the Black Hole escape velocity equation is finding the radius of the body, provided you know its mass. Squaring both sides of the equation and rearranging the items:

$$c^2 < 2GM/R$$

$$Rc^2 < 2GM$$

This results in the equation:

$$R < 2GM/c^2$$

Substituting in values for **G** and **c**, you get:

$$R < 2*(6.674*10^{-20} \text{ km}^3/\text{kg-s}^2)*M/(2.997*10^5 \text{ km/s})^2$$

$$R < (1.486*10^{-30})*M \text{ km}$$

Thus, given the mass, you can find the required radius for the body to be a Black Hole.

Relativity and Schwarzschild radius

Although, our escape velocity equation for a Black Hole given is based on the classical equations and not the relativistic, it is still valid.

In 1916, scientist Karl Schwarzschild derived what is called the *Schwarzschild radius* from Einstein's gravitational field equations in the General Theory of Relativity.

The Schwarzschild radius represents the event horizon of a Black Hole or the limiting radius where nothing can leave. Its equation is:

$$R_e = 2GM/c^2$$

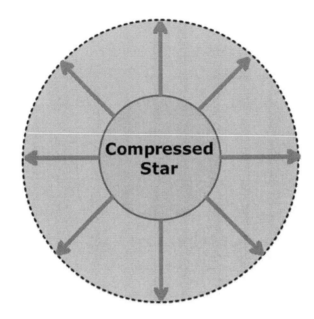

Black Hole and Schwarzschild radius

In the chapter on *Gravitational Escape Velocity*, it was shown that at greater altitudes, the required escape velocity decreases. The Schwarzschild radius represents a limiting separation from the center for light to escape from a Black Hole.

At separations less than this radius, the required escape velocity is *greater* than the speed of light.

Black Hole with mass of our Sun

An application of the event horizon equation is if the mass of the black hole equaled the mass of our Sun ($1.988*10^{30}$ kg). In that case, its Schwarzschild radius or event horizon would be:

$$\mathbf{R_S} = (1.486*10^{-30})*(1.988*10^{30}) \text{ km} = 2.954 \text{ km}$$

In other words, a star with the mass of our Sun with its matter compressed to a radius of less than 2.954 km would be a Black Hole, because the escape velocity would be greater than the speed of light.

Summary

A Black Hole has a gravitational field is so strong than not even light can escape its pull. The gravitational escape velocity equation for a Black Hole substitutes the speed of light for the velocity.

The event horizon or Schwarzschild radius is the defining size of a Black Hole and can be determined from the escape velocity equation. You can use the equation to calculate the size of the Sun if its mass were compressed enough to make it a Black Hole.

The escape velocity equation for a Black Hole is:

$$s_e = \sqrt{(2GM/R)} > c$$

The Schwarzschild radius is:

$$R_S = 2GM/c^2$$

Mini-quiz to check your understanding

1. Why can't particles going faster than the speed of light escape from a Black Hole?

    a. They can escape; only light can't escape

    b. There are no particles in a Black Hole

    c. Matter cannot exceed the speed of light

2. What does the Schwarzschild radius equation prove?

    a. Schwarzschild was smarter than Einstein

    b. The escape velocity equation fits both classical and relativistic theories

    c. No one is really sure what it proves

3. What would happen if the radius of the Sun was compressed to 6 km?

    a. It would not be a Black Hole

    b. The Earth would be the size of a peanut

    c. The speed of light would change

Answers

1c, 2b, 3a

Summary

You have learned much about gravity and gravitation from reading this book.

Gravity

You've learned that gravity is an approximation of gravitation that applies to objects relatively close to the Earth. There is a similar type of gravity for the Moon, planets and Sun.

The equation for the force of gravity is: $\mathbf{F} = \mathbf{mg}$, where $\mathbf{g}$ is the constant acceleration due to gravity.

Since $\mathbf{g}$ is a constant acceleration, you have seen that all objects fall at the same rate, irrespective of mass. Also, the rate an object falls is independent of its horizontal velocity.

Starting with the fact that the acceleration due to gravity is a constant, you have seen how to derive the velocity, distance and time equations for falling objects, as well as those projected upward or downward.

You then saw how you can use those equations in a number of applications, including finding the distance of an artillery projectile and the rocket velocity necessary to escape from Earth. You also looked at Newton's cannon and the possibilities of shooting a cannonball around the Earth, as well as how to create artificial gravity.

Other applications include doing work against gravity and how gravity can do work against inertia.

Gravitation

The book then looks at gravitation and the force of attraction between all bodies of mass. You saw how the gravity equation came from the gravitation equation.

You have also seen that there are several theories about what gravitation is and how it works, starting with the laws of gravitation defined by Isaac Newton, continuing for large bodies and velocities with Albert Einstein's General Relativity Theory of Gravitation and then looking at the atomic level with the Quantum Gravitation Theory.

Applications of gravitational concepts that you learned about in this book include the influence of gravitation among stars and planets in the Universe, gravitational attraction and center of mass, orbits, how dark matter may affect gravitation, escape velocity and effect of gravitation in Black Holes.

More to learn

There is much more that you can learn about gravity and gravitation. There are physicists and scientists who have devoted their lives to the study of the concepts, theories and applications.

Hopefully this material has been useful in your understanding of gravity and gravitation principles, as well as to help and inspire you in your future scientific studies.

Resources

The following books and websites are references and resources on gravity and gravitation.

Books

Physical Science textbooks

Conceptual Physical Science
Chapter 4 Gravity and Satellite Motion
Paul G. Hewitt, John Suchocki and Leslie A. Hewitt;
Addison-Wesley Longman (2004)
ISBN 0321203925

Physical Science: A Unified Approach
Chapter 5 Motion and Gravity
Jerry Schad, San Diego Mesa College;
Brooks/Cole Publishing Co. (1995)
ISBN 0534192483

Physics textbooks

Physics: Algebra/Trig
Chapter 5 Centripetal Force and Gravity
Eugene Hecht, Adelphi University;
Brooks/Cole Publishing (2002)
ISBN 0534377297

Physics: Principles with Applications
Chapter 5 Circular Motion and Gravitation
Douglas C. Giancoli; Prentice Hall (2004)
ISBN 0130606200

Gravity books

Reinventing Gravity: A Physicist Goes Beyond Einstein
John W. Moffat; Smithsonian Publishing (2008)
ISBN 0061170887

Head First Physics: A learner's companion to mechanics and practical physics
Heather Lang; O'Reilly Media (2008)
ISBN 0596102372

Gravity: An Introduction to Einstein's General Relativity
J. B. Hartle; Benjamin Cummings Publishing (2003)
ISBN 0805386629

Gravitation books

Gravitation (Physics Series)
Charles W. Misner, Kip S. Thorne, John Archibald Wheeler, and John Wheeler; W. H. Freeman Publishing (1973)
ISBN 0716703440

Gravitation and Inertia
Ignazio Ciufolini and John Archibald Wheeler;
Princeton University Press (1995)
ISBN 0691033234

Principles of Cosmology and Gravitation
Michael V. Berry; Taylor & Francis Publishing (1989)
ISBN 0852740379

The Gravitational Constant: Generalized Gravitational Theories and Experiments
Venzo De Sabbata, George T. Gillies and Vitaly N. Melnikov;
Springer Publishing (2004)
ISBN 1402019556

Spacetime and Geometry: An Introduction to General Relativity
Sean Carroll; Benjamin Cummings Publishing (2003)
ISBN 0805387323

Websites on Gravity

Gravity basics

Force of Gravity - Universe Today Magazine
http://www.universetoday.com/guide-to-space/physics/
force-of-gravity/

Earth's gravity - Wikipedia
http://en.wikipedia.org/wiki/Earth's_gravity

How does gravity work? - How Stuff Works
http://www.howstuffworks.com/question232.htm

Standard gravity - Average value, as compared to variation due
to position on Earth - Wikipedia
http://en.wikipedia.org/wiki/Standard_gravity

International Gravity Formula - Variation of gravity with distance from equator - Geophysics dept. University of Oklahoma
http://geophysics.ou.edu/solid_earth/notes/potential/igf.htm

I feel 'lighter' when up a mountain but am I?
National Physics Laboratory FAQ
http://www.npl.co.uk/reference/faqs/i-feel-'lighter'-when-up-a-
mountain-but-am-i-(faq-mass-and-density)

Vectors

Vectors - Fundamentals and Operations - Physics Classroom
http://www.physicsclassroom.com/Class/vectors/u3l1a.cfm

Basic Vector Operations - HyperPhysics
http://hyperphysics.phy-astr.gsu.edu/hbase/vect.html

Vectors and Direction - Physics Classroom
http://www.physicsclassroom.com/Class/vectors/u3l1a.cfm

Gravity constant

Acceleration Due to Gravity - TutorVista.com
http://physics.tutorvista.com/motion/acceleration-due-to-gravity.html

The Value of g - Physics Classroom
http://www.physicsclassroom.com/class/circles/u6l3e.cfm

Acceleration Due to Gravity - Haverford College
http://www.haverford.edu/educ/knight-booklet/accelarator.htm

The Acceleration of Gravity - Physics Classroom
http://www.physicsclassroom.com/class/1dkin/u1l5b.cfm

Difference Between Pound-Force and Pound-Mass - Engineerography Blog
http://engineerography.com/2009/03/what-the-hecks-the-difference-between-pound-force-and-pound-mass/

Weak Equivalence Principle

Test of Weak Equivalence Principle
Harvard-Smithsonian Center for Astrophysics
http://www.cfa.harvard.edu/PAG/index_files/Page1098.htm

Feather & Hammer Drop on Moon -YouTube video
http://www.youtube.com/
watch?v=5C5_dOEyAfk&feature=related

Falling bodies

Falling Bodies - Physics Hypertextbook
http://physics.info/falling/

Equations for a falling body - Wikipedia
http://en.wikipedia.org/wiki/Equations_for_a_falling_body

Gravity Calculations for Earth - Online Calculators
http://www.gravitycalc.com/

Kinematic Equations and Free Fall - Physics Classroom http://www.physicsclassroom.com/class/1dkin/u1l6c.cfm

Work and gravity

Work by gravity by Sunil Kumar Singh - Connexions
http://cnx.org/content/m14101/latest/

Gravity and Inertia in Running -Locomotion and Biology paper
http://jeb.biologists.org/cgi/reprint/203/2/229.pdf

Artificial gravity

Artificial Gravity and the Architecture of Orbital Habitats
Theodore W. Hall - Space Future; detailed technical paper
http://www.spacefuture.com/archive/artificial_gravity_and_the_
architecture_of_orbital_habitats.shtml

SpinCalc: an artificial-gravity calculator in JavaScript
http://www.artificial-gravity.com/sw/SpinCalc/SpinCalc.htm

Artificial Gravity - Technical resources from Theodore W. Hall
http://www.artificial-gravity.com/

Artificial gravity - Wikipedia
http://en.wikipedia.org/wiki/Artificial_gravity

The Physics of Artificial Gravity - Popular Science magazine
http://www.popsci.com/scitech/article/2009-02/
physics-artificial-gravity-part-two

Simulated Gravity with Centripetal Force
Oswego City School District Exam Prep Center, New York
http://regentsprep.org/regents/physics/phys06/bartgrav/
default.htm

Newton's Cannon

Newton's Cannonball and the Speed of Orbiting Objects
Bucknell University Physics
http://www.eg.bucknell.edu/physics/astronomy/astr101/
specials/newtscannon.html

Newton's Orbital Cannon - Encyclopedia Astronautica
http://www.astronautix.com/lvs/newannon.htm

Newton's cannonball - Wikipedia
http://en.wikipedia.org/wiki/Newton's_cannonball

Interactive Demonstration of Newton's Cannon
http://waowen.screaming.net/revision/force&motion/ncananim.
htm

Artillery projectile

Trajectories - HyperPhysics
http://hyperphysics.phy-astr.gsu.edu/hbase/traj.html

Projectile Motion - Wikipedia
http://en.wikipedia.org/wiki/Projectile_motion

Trajectory for Projectile Motion
John Hopkins University - Deprtment of Physics & Astronomy
http://en.wikipedia.org/wiki/Trajectory_of_a_projectile

Trajectory of a projectile - Wikipedia
http://en.wikipedia.org/wiki/Trajectory_of_a_projectile

Characteristics of a Projectile's Trajectory - Physics Classroom
http://www.physicsclassroom.com/class/vectors/u3l2b.cfm

Websites on Gravitation

Gravitation - NASA Worldbook
http://www.nasa.gov/worldbook/gravitation_worldbook.html

Gravitation - Wikipedia
http://en.wikipedia.org/wiki/Gravitation

Classical theories of gravitation

Classical theories of gravitation - Wikipedia
http://en.wikipedia.org/wiki/Classical_theories_of_gravitation

Sir Isaac Newton: The Universal Law of Gravitation - Dept. of
Physics & Astronomy University of Tennessee-Knoxville
http://csep10.phys.utk.edu/astr161/lect/history/newtongrav.
html

Newton's Law of Universal Gravitation - SparkNotes SAT
http://www.sparknotes.com/testprep/books/sat2/physics/chapter11section2.rhtml

Newton's Law of Universal Gravitation - Wikipedia
http://en.wikipedia.org/wiki/Newton's_law_of_universal_gravitation

Newton's Law of Universal Gravitation - Physics Classroom
http://www.physicsclassroom.com/class/circles/u6l3c.cfm

Gravitational Acceleration - Wikipedia
http://en.wikipedia.org/wiki/Gravitational_acceleration

Gravitational Acceleration Calculator
http://www.ajdesigner.com/phpgravity/gravity_acceleration_equation.php

Gravitational Acceleration Tutorial and Calculators -
EasyCalculation.com
http://www.easycalculation.com/physics/classical-physics/learn-gravitational-acceleration.php

History of Theories of Gravitation - University of St. Andrews,
Scotland - http://www-groups.dcs.st-and.ac.uk/~history/Hist-Topics/Gravitation.html

Relativity and gravitation

Gravitation and the General Theory of Relativity -
University of Tennessee at Knoxville
http://csep10.phys.utk.edu/astr162/lect/cosmology/gravity.html

Einstein's Theory of Relativity versus Classical Mechanics -
Complete online book by Paul Marmet
http://www.newtonphysics.on.ca/einstein/index.html

General Theory of Relativity - Wikipedia
http://en.wikipedia.org/wiki/General_theory_of_relativity

General relativity - Overview from Einstein Online
http://www.einstein-online.info/spotlights/gr

Tests of general relativity - Wikipedia
http://en.wikipedia.org/wiki/Tests_of_general_relativity

The Principle of Equivalence from Einstein's Theory of Relativity versus Classical Mechanics - online book by Paul Marmet - http://www.newtonphysics.on.ca/einstein/chapter10.html

Equivalence principle (relativity) - Wikipedia
http://en.wikipedia.org/wiki/Equivalence_principle

Quantum gravitation

Graviton - - Wikipedia
http://en.wikipedia.org/wiki/Graviton

Gravitational waves - Overview from Einstein Online
http://www.einstein-online.info/spotlights/gravWav

String theory - Wikipedia
http://en.wikipedia.org/wiki/String_theory

What is String Theory? - About.com Physics - http://physics.about.com/od/quantumphysics/f/stringtheory.htm

Quantum mechanics - Wikipedia
http://en.wikipedia.org/wiki/Quantum_mechanics

What is Quantum Gravity? - About.com Physics
http://physics.about.com/od/quantumphysics/f/quantumgravity.htm

Quantum Gravity - Einstein Online
http://www.einstein-online.info/spotlights/background_independence

Loop quantum gravity - Wikipedia
http://en.wikipedia.org/wiki/Loop_quantum_gravity

What is Loop Quantum Gravity? - About.com
http://physics.about.com/od/unifiedfieldtheories/f/loopquant-grav.htm

Finding gravitational constant

Gravitational Constant - Wikipedia
http://en.wikipedia.org/wiki/Gravitational_constant

Controversy over Newton's Gravitational Constant -
University of Washington
http://mist.npl.washington.edu/eotwash/experiments/bigG/
bigG.html

Final Demystification of the gravitational constant variation - Blaze Labs Research; extensive research on variations and Unified Theory studies
http://www.blazelabs.com/f-u-massvariation.asp

Cavendish Experiment - Wikipedia
http://en.wikipedia.org/wiki/Cavendish_experiment

Cavendish Experiment - Harvard University Physics
http://www.fas.harvard.edu/~scdiroff/lds/NewtonianMechanics/
CavendishExperiment/CavendishExperiment.html

Measurement of Gravitational Constant (PDF) - Rice Univ.
http://www.owlnet.rice.edu/~dodds/Files332/cavendish.pdf

Henry Cavendish - Weighing the Earth
http://www.juliantrubin.com/bigten/cavendishg.html

Bending Spacetime in the Basement: Measuring Gravitation Homemade graviational experiments - John Walker of Fourmi-lab Switzerland; Some controversy on validity of results - http://www.fourmilab.ch/gravitation/foobar/

Escape velocity

Escape velocity - HyperPhysics
http://hyperphysics.phy-astr.gsu.edu/hbase/vesc.html

Gravitational Potential - Wikipedia
http://en.wikipedia.org/wiki/Gravitational_potential

Gravitational Potential - A-level Physics - WikiBooks
http://en.wikibooks.org/wiki/A-level_Physics_%28Advancing_
Physics%29/Gravitational_Potential

Gravitational Potential Energy - - SparkNotes SAT Physics
http://www.sparknotes.com/testprep/books/sat2/physics/chap-
ter11section3.rhtml

Gravitational or Gravity Well - Wikipedia
http://en.wikipedia.org/wiki/Gravitational_well

Black Holes

Black Holes - NASA Astrophysics
http://science.nasa.gov/astrophysics/focus-areas/black-holes/

Black Holes - Overview from Einstein Online
http://www.einstein-online.info/spotlights/blackHoles

The Schwarzschild Radius - HyperPhysics
http://hyperphysics.phy-astr.gsu.edu/hbase/astro/blkhol.
html#c2

The Schwarzschild Radius: Nature's Breaking Point - NOVA
- http://www.pbs.org/wgbh/nova/physics/blog/2011/12/the-
schwarzschild-radius-natures-breaking-point/

Gravitational singularity - Wikipedia
http://en.wikipedia.org/wiki/Gravitational_singularity

Singularities and Black Holes - Stanford University
http://plato.stanford.edu/entries/spacetime-singularities/

Index

Symbols

2001: A Space Odyssey 245

A

Acceleration due to gravity 32, 264
Albert Einstein 243, 273, 287
Algebra 23
Apollo Moon mission 491
Applications
  Gravitation 341
    Cavendish experiment 349
    Electrostatic forces 323
    Equivalence principles 315
    Force, two objects 343
    Potential energy 333
    Speed 327
  Gravity 179
    Artificial gravity 243
    Artillery projectile 215
    Center of gravity 251
    Escape velocity 235
    Newton's cannon 227
    Potential energy 181
    Sideways motion 209
    Work against gravity 199
    Work against inertia 189
Artificial gravity 243
Artillery projectile 215, 518
Author 529

B

Balance scale 60
Big Bang 357
Black Hole 303, 505, 522

C

Calculus 23, 75, 81, 82, 87, 284
Cannonball, Newton's 227
Cartesian coordinate system 38, 44
Cavendish experiment 349
Cavendish, Henry 281, 349
Center of gravity 251
Center of mass 371, 373
  Acceleration 390
  Components 397
  Derivation 421
  Gravitational motion 379, 387
  Location 387
  Radial motion 403
  Radial vector 399
  Relative motion 393
  Tangential motion 411
  Tangential vector 400
  Velocity 389
Centrifugal force 244, 424
Circular orbit 464
  Cannonball 230
  Planetary 443, 453
Classical theories of gravitation 518
Components, vectors 38, 375, 397

Conventions 23
  Exponents 24
  Multiplication sign 23
  Units 25
  Vectors 43
Coordinate system
  Gravitation 438
  Gravity 437
  Gravity equations 38
Coulomb's Law 324

D

Dark energy 299, 302
Dark matter 291, 299
Definitions
  Center of gravity 251
  Center of mass 379
Deflection of light rays 278
Density of Earth 349
Derivations 73
  Gravitation
    Escape velocity 483
    Gravitational constant 351
    Length of year 453
    Orbiting center of mass 421
    Potential energy at infinity 335
  Gravity
    Artillery displacement 218
    Displacement-Time 87
    Displacement-Velocity 93
    Gravity constant 49
    Velocity-Time 81
Derivative 75, 81, 88
Direction convention
  Gravitational motion 437
  Gravity equations 43
Displacement 37
  Derivations
    Displacement-Time 87
    Displacement-Velocity 93
  Equations 101, 123, 148
    Falling objects 111
    Projected downward 133
    Projected upward 159
  Vector definition 44
Distance

Displacement 40
  Total 164
Doppler effect 301
Double stars 425

E

Earth
  Escape velocity 238
  Mass 499
  Orbit around Sun 446
Earth bulge 358
Einstein, Albert 243, 263, 287
Einstein Equivalence Principle 319
Einstein Field Equations 325
Electromagnetic force 308
Electrostatic forces 323
Elliptical orbits 229
  Large 466
  Small 465
Energy
  Dark 299
  Falling objects 181
  Objects projected downward 183
  Objects projected upward 185
Equations
  Falling objects 99, 101
    Displacement 111
    Time 115
    Velocity 105
  Projected downward 121, 123
    Displacement 133
    Time 139
    Velocity 127
  Projected upward 145, 147
    Displacement 159
    Time 169
    Velocity 151
Equivalence principle
  Experimental verification 57
  Gravity 55
  Strong 319
  Types of mass 318
  Weak 315
Escape velocity
  Black Hole 505
  Derivation 483

Gravity and Gravitation

Effect of Sun on Earth 499
Equation 473, 487
Escape from Moon 479
Escape from Sun 479
Examples 478
Gravitational 471, 473
Gravity 235
Potential energy 337
References 521
Saturn V rocket 491
Exponents, convention 24

F

Falling objects
Displacement 111
Equations 99, 101
Time 115
Velocity 105
Force of attraction
Boy and girl 344
Earth and Moon 343
Girl and Moon 345
Free fall principle 55
Fundamental forces 294, 307

G

Galaxies, formation 357
Galileo 57, 275
General Theory of Relativity 243,
    270, 287, 296
Geometric visualization 39
Gravitation 22, 259
Applications 341
Coordinate system 438
Direction convention 437
Effect of dark matter 299
Electrostatic forces 323
Equivalence principles 315
Escape Velocity 473
Force 277
Fundamental force 307
Overview 261
Principles 313
Properties 261
Quantum Theory 293

Speed 278
Theories 262, 267
Tides on Earth 363
Universe 357
Gravitational
Constant 521
Force 309, 343
Mass 59, 319
Potential energy 333
Speed 289, 327
Gravitomagnetism 325
Graviton 295, 327
Gravity 21, 27
Acceleration constant 55
Applications 179
Artificial gravity 243
Artillery projectile 215
Center of gravity 251
Coordinate system 437
Equivalence principle 55
Escape velocity 235
Force equation 31, 264
Gravity constant 49
Horizontal motion 67
Inertia and gravity 199
Mass and weight 59
Moon 33
Newton's cannon 227
Potential energy 181
Sideways motion 209
Variation with altitude 52
Vectors 37
Work and gravity 189, 199
Gravity constant 516

H

Henry Cavendish 281, 349
High tide 364
Hofava gravitation 291
Hooke, Robert 275
Horizontal displacement 45
Horizontal motion 67, 209
Howitzer 215
Hyperbolic path 415, 467

Index

I

Inertia 189
Inertial mass 61, 318
Infinite speed 328
Integral sign 82, 89
Integrate 75
International Space Station 243
Isaac Newton 227, 275, 281, 349

J

Jupiter orbits Sun 448

K

Kepler's Laws 275, 328
Kinetic energy 182, 336, 485
Knowledge needed 23
Kurtus, Ron 529

L

Law of Universal Gravitation 31,
    269, 275, 287, 327
Law opposed to theory 269
Laws of Motion 275
Leaning Tower of Pisa 57
Length of year 453
Le Verier, Urbain 328
Light, deflection 290
Location, center of mass 387
Loop Quantum Gravity Theory 296
Luminosity 300

M

Magnitude, vectors 38
Mass
  Earth 499
  Gravitational 59
  Inertial 61
  Measuring 60
  Sun 500
  Unit 59
  Weight 59
Mass, Center of 371
Matter, dark 299

Maximum displacement
  Artillery 218
  Time 162, 170, 172
  Velocity 154, 160
Maxwell's field equations 325
Measuring
  Gravitational constant 350
  Mass 60
    Doppler effect 300
    Luminosity 300
  Weight 64
Mercury, orbit 289, 328
Moon, escape velocity 239
Moon orbits Earth 360, 443
Mortar artillery 222
Motion projected at angle 70
Multiplication, convention 23

N

NASA space program 491
Newton, Isaac 269, 275, 281, 349
Newton's Cannon 227, 465, 517
Nuclear forces 307

O

Orbital motion 419
  Cannonball 229
  Circular planetary 443
  Conditions 359
  Length of year 453
  Mercury 289
  Period 455
  Relative to other object 431
  Velocity 453, 463
Oscillation period 353
Overviews
  Center of mass 373
  Escape velocity 473
  Falling objects 101
  Force of gravity 31
  Gravitation 261
  Gravity equation derivations 75
  Objects projected downward 123
  Objects projected upward 147
  Theories of gravitation 269

Gravity and Gravitation

P

Parabolic path 464, 466
Planets, formation 357
Potential energy 484
  Gravitational 333
  Gravity 181
Principia Mathematica 275
Principles, Gravitation 313
Projected downward
  Displacement 133
  Equations 121, 123
  Time 139
  Velocity 127
Projected upward
  Displacement 159
  Equations 145, 147
  Time 169
  Velocity 151
Projectile path 217
Properties of gravitation 261
Purposes of book 23
Pythagorean Theorem 39, 69

Q

Quadratic equation 89
Quantum theories 271
  Gravitation 293, 327, 520
  Mechanics 293

R

Radial motion
  Component 375, 399
  Effect of tangential 407
  Expansion of Universe 406
  Gravitational 403
Radius of black hole 506
Red-shift 290
Relative motion 393
Relativity 519
  Deflection of light 290
  General theory 287
  Special theory 287
Resources
  Books 513

Websites 515
Ron Kurtus 529

S

Saturn V rocket 491
Scalars
  Gravitation 439
  Gravity 40
School for Champions 529, 531
Schwarzschild radius 505
Scientific
  Method 25
  Notation 24
Shooting bullet 212
Sideways Motion 209
Significant figures 24
Space Station 243
Spacetime 288
Special Theory of Relativity 329
Speed of gravitation 289, 327
Speed versus velocity 40
Stars, formation 357
String Theory 295, 307, 329
Strong Equivalence Principle 319
Strong force 307
Summary of book 511
Sun
  Escape velocity 239
  Mass 500, 508

T

Tangential motion 411
  Component 375, 400
  Effect of radial velocity 415
  Orbits 431
Theories
  General Relativity 270, 287, 327
  Gravitation 262, 267
  Quantum theory 293
  Special Relativity 287
  Standard model 307
  String 307
  Tidal configuration 366
Thought experiment 227
Tides on Earth

From Moon 363
From Moon and Sun 367
Time
  Derivations
    Displacement-Time 87
    Velocity-Time 81
  Time equations 102, 124, 148
    Falling objects 115
    Maximum displacement 152, 170
    Objects projected downward 139
    Objects projected upward 169
Torsion
  Balance 349
  Coefficient 352
Total distance 164
Total work 206
Translunar injection 492

U

Unit of mass 59
Units, convention 25
Universal Gravitation
  Constant 281, 454, 484, 499
  Equation 264, 270, 277, 281, 343, 349
  Law 269, 275
Universe 357

V

Vectors 515
  Components 38
  Convention for direction 37
  Displacement 44
  Explanation 68
  Gravitation 439
  Gravity direction 43
  Horizontal velocity 68
  Velocity 45
Velocity
  Components 215
  Derivations
    Displacement-Velocity 93
    Velocity-Time 81
  Equations 147
  Escape velocity 471

Black Hole 505
  Derivation 483
  Effect of Sun 499
  Gravitational 473
  Saturn V Rocket 491
  Velocity equations 101, 123
    Falling objects 105
    Projected downward 127
    Projected upward 151
Velocity, center of mass 389
Vertical
  Displacement 44
Vectors 37

W

Wave-particle duality 272, 294
Weak equivalence principle 55, 315, 516
Weak force 308
Weight
  Mass 59
  Measuring 64
  On the Moon 65
Weightlessness 243
Work
  Against gravity 200
  Against inertia 189, 201
  As change in energy 191
  As force times distance 189
  By gravity 189
  Definition 199
  Ending at zero velocity 204
  Projecting object upward 195, 202

Z

Zwiky, Fritz 299

About the Author

Ron Kurtus

Ron Kurtus has a broad scientific and mathematical background, which includes working as an electro-optical systems engineer and engineering manager for the U.S. Air Force Space Defense programs and developing weather satellites at the Santa Barbara Research Center.

He also taught Mathematics at the University of Missouri, Physical Science at the Milwaukee Area Technical College and Mathematics at the Waukesha County Technical College.

Kurtus is the founder and author of the award winning *School for Champions* educational website, which has been cited as a resource in over 130 published books from thirteen countries. The website is found at: *www.school-for-champions.com*.

He is also author of the book *Tricks for Good Grades: Strategies to Succeed in School*.

Ron lives with his wife Eleanor and Golden Retriever Zoe in Lake Oswego, Oregon.

School for Champions

Ron Kurtus' *School for Champions* is an award-winning educational website that is used by over 600,000 students each month.

The website not only provides educational lessons but also explains how users can advance in their careers and businesses, as well as to advocate excellence and fine character. A major goal of the school is to encourage people to help others and champion worthy causes.

The material is in this book is based on the Gravity and Gravitation lessons in the Physics section of the website.

This book is published by SfC Publishing Co., which is the media arm of the School for Champions.

You can access the School for Champions website at:

http://www.school-for-champions.com

23804888R00293

Printed in Great Britain
by Amazon